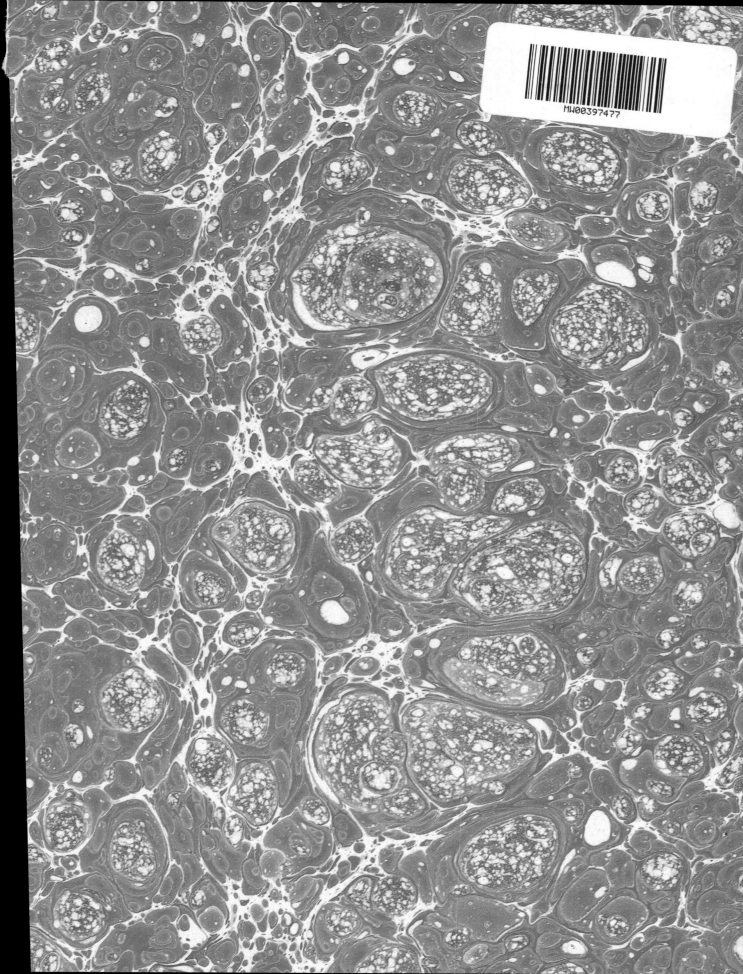

THE HG PANZER DIVISION

by
ALFRED OTTE

Schiffer Publishing Ltd

1469 Morstein Road
West Chester, Pennsylvania 19380

List of Waffen-SS Ranks and their World War 2 German and US Equivalents

Waffen SS	German WW 2 Army	US WW 2 Army
General Officers		
—no equivalent—	Generalfeldmarschall	General of the Army
Oberstgruppenführer	Generaloberst	General
Obergruppenführer	General	Lieutenant General
Gruppenführer	Generalleutnant	Major General
Brigadeführer	Generalmajor	Brigadier General
Staff Officers		
Oberführer	—no equivalent—	—no equivalent—
	(wore the shoulder strap of a colonel)	
Standartenführer	Oberst	Colonel
Obersturmbannführer	Oberstleutnant	Lieutenant Colonel
Sturmbannführer	Major	Major
Company Officers		
Hauptsturmführer	Hauptmann	Captain
Obersturmführer	Oberleutnant	1st Lieutenant
Untersturmführer	Leutnant	2nd Lieutenant
Officer Candidates (basically equal to Oberfeldwebel & Feldwebel)		
Oberjunker	Oberfähnrich	—no equivalent—
Junker	Fähnrich	—no equivalent—
Noncommissioned Officers		
Sturmscharführer	Stabsfeldwebel	Sergeant Major
Oberscharführer	Oberfeldwebel	Master Sergeant
Scharführer	Feldwebel	Technical Sergeant
Unterscharführer	Unterfeldwebel	Staff Sergeant
	Unteroffizier	Sergeant
Enlisted Men		
—no equivalent—	Stabsgefreiter	Admin Corporal
Rottenführer	Obergefreiter	Corporal
Sturmmann	Gefreiter	Corporal
SS-Obersoldt *	Obersoldt *	Private 1st Class
SS-Soldat *	Soldat *	Private

*Note: "Soldat" is a general term. Other words used here are Schütze, Grenadier, Füsilier, depending upon the combat arm to which the soldier belonged.
  Source of US World War 2 army equivalents: War Department Technical Manual TM-E 30-451 *Handbook on German Military Forces*, 15 March 1945.

Originally published under the title, "Die Weissen Spiegel", by Podzun-Pallas-Verlag, copyright Podzun-Pallas-Verlag, 6360 Friedberg 3 (Dorheim), © West Germany, ISBN: 3-7909-0185-7.

Copyright © 1989 by Schiffer Publishing.
Library of Congress Catalog Number: 89-063370.

Printed in the United States of America.
ISBN: 0-88740-206-2

Published by Schiffer Publishing, Ltd.
1469 Morstein Road
West Chester, Pennsylvania 19380
Please write for a free catalog.
This book may be purchased from the publisher.
Please include $2.00 postage.
Try your bookstore first.

CONTENTS

INTRODUCTION

In October, during the heavy fighting at the Weichsel, the Fallschirmpanzerkorps HG (Parachute tank corps), was formed from the Fallschirm-Panzer-Division HG and other units. As a part of the Luftwaffe (air force) it belonged to the Fallschirmtruppe (parachute troops), and reported to the Fallschirm-Oberarmeekommando 1 (parachute army high command). It was used in the same way as other units of the parachute troops which, different from today, were part of the air force, within the framework of the army.

The Fallschirmpanzerkorps HG and its previous formations have, for many years, been the subject of much interest among the circles of former and present day soldiers. Surprisingly, this interest is particularly high in other countries, especially in Great Britain and the United States, which has brought about many mutual contacts between the veterans' associations and has laid ground to many friendships. The saying "Yesterday's enemies-Today's friends" has become a reality.

This book is meant to be a document of memory for the survivors of the more than 60,000 soldiers who wore the white patches, as well as their families. For the others, it is meant to contribute to the understanding of this special troop.

After the war, numerous and repeated misjudgment was caused by the designation in connection with the name of Göring. To prevent such misinterpretation here, it must be noted that all units were given the use of the then Prussian prime minister and, later, supreme commander of the air force, were units as all others. To attach to them or their members a particular political attitude or political activity disregards the facts and would be unjust. For reasons of historical fact, the units in this book are referred to by their name at that time, historically correct, with GG for General Göring and HG for Hermann Göring. Correspondingly, the abbreviations of the time are used here: RGG for Regiment General Göring and RHG for Regiment Hermann Göring.

The creation of this book was possible only because many comrades of all ranks have volunteered to provide the author with photos and information. They all deserve thanks. Many of the photos were taken by the contributors and are here shown publicly for the first time. As these were taken by amateurs, often under adverse conditions, they reflect the conditions of the time in an unadulterated way.

Alfred Otte

Photos, documents and drawings were provided by the ladies and gentlemen listed below, as well as archives and service organizations. Our thanks to them must be repeated here. The author contributed sixty-five pictures, documents and drawings.

Achstetter, Alfred -Albrecht, Jans-Jürgen-Altenschmidt, Hans-Arnold, Dr. Alfred-Bergold, Elisabeth-Bidinger, Joachim-Böhmler, Rudi-Bongartz, Jakob-Brunsmeier, Walter-Busch, Erich-Dieke, Walter-Englert, Karl-Fallowe, Wilhelm-Fetzer, Rudolf-Fitz, Marie-Flothow, Rolf-Funck, Helmut-Grau, Hans-Joachim-Gernand, Willi - Gross, Herbert-Hagendorfer, Helmut-Hasselbach-Häussler, Wilhelm-Henne, Georg-Hermann, Ernst-von Hippel - alowski, Wilhelm-Kanert, Bruno-Keicher, Wilhelm -Keller, Theodor-Kretschmer, Gustav-Kulp, Amalie -Küstner, Georg-Lange, Walter-Lorch, Franz -Lübberstedt, Heinrich-Lutz, Ernst-Marwan-Schlosser, Rudolf-Metzler, Gerhard-Meyer-Schewe, Friederich-Nünninghof, Heinz W.-Nüske, Hermann-Obermaier, Ernst-Olfermann, Gerhard-Oqueka, Walter-Otte, Karl-Heinz -Peterson, George E.-Ritterbecks, Bert-Roehle, Helmut -Schade, Wilhelm-Scheid, Gerda-Schier, Hans-Schlegel, Ilse-Schlenzig, Ernst-Schmalz, Wilhelm-Schorn, Peter -Seeger, Hubert-Seehawer, Egon-Sextro, Franz-Josef-Spitzbarth, Dr. Reimer-Stuckmann, Wilhelm-Stumhofer, Josef-Thunemann, Willi-Timme, Hans A.-Ueckert, Hans-Weidemann, Willy-Welmering, Bernhard-Wessler, Wilhelm

FROM POLICE DETACHMENT TO LUFTWAFFE REGIMENT

The roots of the RGG (Regiment General Göring) reach back to a police detachment. By decree of February 23, 1933, the then prime minister of Prussia ordered the creation of a motorized police detachment of 418 members, for which outstanding volunteers of the Prussian security police, in particular those from Greater Berlin, were to be trained to full police status. Major Wecke of the security police was named commander of the detachment. The new police unit was given its designation after the name of its commander, Police Department Wecke (z.b.V., for special requirements).

The detachment was put together a few days later in the Augustaner barracks in the Friesenstreet of the Kreuzberg quarter of Berlin. It was divided into a detachment command, three police squads, the communications platoon and the motorcycle platoon. Later two special police vehicles were added. These were armored and carried mounted machine-guns, but were quite unwieldy vehicles.

The original mandate of pure police functions shifted within months, at an increasing pace towards the military area which brought about a change in equipment and grouping. Thus, even in May 1933 the original detachment evolved into the Police Group Wecke z.b.V. This group was removed, on July 17, 1933, from the general security police, made the direct responsibility of the Prussian prime minister and was thus renamed State Police Group Weche z.b.V. The group was the first state police force in Germany and, from that point onward, became the example which all other state police forces in Germany used for their development. At the same time, the blue police uniforms were replaced by a new green uniform, in color and cut similar to that of the German army. Soon after, the subdivisions and ranks were made equal to those of the army: 'Hundertschaft' (troop of 100 men) company, detachment battalion, group regiment, constable corporal, first constable sergeant, etc.

The State Police Group Wecke z.b.V. was renamed, on January 12, 1934, State Police Group GG and the First Detachment, as the original one, was now known as (I) Detachment Wecke-State Police Group GG. At the same time all members of the group were provided with a sleeve band "L.P.G. General Göring" (Landes Polizei Gruppe General Göring) to be worn on the left foresleeve.

The training, from now on exclusively military, followed the army service manual and was intensified in this through the mutual exchange with army officers and non-commissioned officers, as well as courses at training facilities of the army and ongoing maneuvers in the area of Mark Brandenburg and, finally, through periods of time spent at various army training grounds.

In July 1933, the State Police Group Wecke z.b.V. consisted of the group staff, two detachment staffs, eight squads, the cavalry troop, the motorcycle troop, vehicle squad, communications troop nd the musical corps.

On June 6, 1934, Lt. Colonel Jakoby of the state police was made commander of the State Police Group GG. Organization, armament and equipment in April 1935 were equivalent to those of a motorized infantry regiment of the German army.

The legislation concerning the establishment of the German armed forces of March 16, 1935, directed that the German army in peace time, including the integrated police units, be divided into twelve corps commands and 36 divisions. The above mentioned police units were state police groups under the authority of the states. When the integration of these units into the new army took place in summer 1935, the State Police Group GG, as the only police group, was exempted. Since it was the 'house regiment' of the then Prussian prime minister Göring, whose name it used, the group was integrated into the newly created air force. Göring, since April 1933 General of the Infantry and since March 1935 General of the Pilots, had in the meantime been appointed supreme commander of the air force.

The State Police Group GG had, at the time of integration into the air force, a strength of 1856 policemen and consisted of the group staff with music corps, the communications platoon, the cavalry platoon, the engineer platoon, the 13th mortar company, the 14th anti-tank company and the 15th company (with engineer platoon and motorcycle platoon), as well as the I., II. and III rifle battalions, each with a communications platoon, three rifle companies and a machine-gun company.

Police accommodation in the Friesenstreet of Berlin Kreuzberg, the former barracks of the Queen-Augusta-Guard-Grenadier-Regiment No. 4. Here the Police Detachment Wecke z.b.V. was born and put together in February 1933 on the third floor, left wing of the building. Recruits of the Prussian security police receiving basic training.

Further training for police duties is provided in the police stations.

The Polizeiabteilung Wecke z.b.V. received two special police vehicles, very unwieldy armored vehicles on solid rubber tires with two heavy machine guns in turning turrets.

Within a few months the police detachment was expanded to police group. It was removed, in July 1933, from the security police and made directly responsible to the Prussian prime minister. It was named "Landespolizeigruppe Wecke z.b.V.". Instead of the blue police uniforms, a green uniform, cut similar to that of the army, was now worn. The changing of the guard is wearing the new uniform for the first time. On the steel helmet, which had replaced the old style helmet (Tschako), the black-white-red insignia is worn on the left side.

On September 13, 1933, the Lanespolizeigruppe Wecke received its colors in the Berlin-Lichterfield barracks where the unit was now quartered. This flag went on to the RGG in 1935 and was carried by its Fallschirm-Schutzen-Battalion (parachute rifle battalion) which took it to Stendal when it was made I./Fallschirm-Jager-Regiment in 1938. It was altogether the first standard a German army unit received after the First World War.

On June 6, 1934 the Colonel of the State Police, Wecke, handed over the LPG GG to his successor, Lt. Colonel of the State Police Jakoby. The complete group is assembled in parade formation for a subsequent march past in the drill grounds of the Lichterfelds barracks. The officer on the far left is Major Schrepffer, commander II. Detachment.

For special occasions the Tschako head-dress continued to be worn. With the completed change of uniform in the Fall of 1933 a fur piece, in memory of the Jagdtruppe (hunter troop) of the old army, was introduced to the Tschako which was very popular. In January 1934, the Landespolizeigruppe Wecke z.b.V. was renamed "Landespolizeigruppe General Göring:. A green sleeve band with the beauty of the Mark Brandenburg. A break in a march by the roadside.

The Prussian state demanded frugality from its officials, including those in the police. During longer maneuvers, quarters were taken up in farms, the nights were spent on straw.

At the end of a large two-week exercise of the **LPG GG** (unofficially already referred to as RGG) in September 1935, which completed the move from the State Police Regiment to the Air Force Regiment, the State of Saxony was included in the exercise terrain. Members of the **LPG GG** at the fortress Konigstein on the Elbe River. The car of the commander of the II. Detachment still carries the Berlin police license plates "IA". For the air force, the move of the **LPG GG** was a great gain. Because of the meticulous screening of the former police recruits and the training directed towards character, knowledge and aptitude, the fledgling air force was able to take over an elite unit from which many sergeants advanced to commissioned officers and not a few former commissioned police officers advanced to General.

Regiment General Göring heute in Jena!

Das Regiment General Göring verläßt in der Frühe des 5. September Berlin, um sich, vollständig motorisiert, auf rund 200 Fahrzeugen zu einer großen Herbstübung nach Thüringen, Bayern und Sachsen zu begeben. An den Uebungen, die unter Leitung des Kommandeurs des Regiments, des Oberstleutnants Jakoby, stattfinden, werden außer den in Berlin benötigten Wachen sämtliche Formationen teilnehmen, also der Regimentsstab mit den unterstellten Formationen und drei Bataillone, zum Teil als Rahmenbataillone. Es handelt sich bei diesen Uebungen, die an den Fahrtagen auf einer

Durchschnittsgeschwindigkeit von 200 Kilometern

aufgebaut sind, um regelrechte Manöver im Regimentsverband. Im Laufe der ersten Tage wird ein Kriegsmarsch unter Fliegersicherung durchgeführt, und zwar in der Annahme, daß das Regiment als Reserve der Obersten Heeresleitung in Berlin verblieben war, in den Nachmittagsstunden des 4. September alarmiert war und zur Verfügung einer an der Mainlinie eingesetzte Heeresgruppe

für den 5. September bis Jena befohlen

wurde. Die Uebungen der nächsten Tage sind darauf zugeschnitten, die Leistungsfähigkeit von Kraftwagen, Material und Mannschaften zu erproben. Die Weiterentwicklung der Lage ist von der Leitung so gedacht, daß das Regiment am 6. September näher an die Kampffront der Mainlinie herangezogen wird, um dann in der Nacht zum 8. September im Raum um Würzburg zum Kampf eingesetzt zu werden. Nach den Rasttagen in Würzburg wird das Regiment am 11. September um 11 Uhr vormittags bei der Eröffnung des Parteikongresses mit den an der Uebung beteiligten Formationen vor seinem Chef General Göring und vor dem Führer an der Luitpoldhalle in Parade stehen. Noch während der Tagung des Parteikongresses sitzt das Regiment auf die Fahrzeuge auf, um am gleichen Tage zu weiteren Uebungen bis Hof

in Bayern zu fahren. Für die fünf Tage vom 11. bis einschließlich zum 15. September ist eine das Höchstmaß der Leistungsfähigkeit beanspruchende Zerreißprobe von Material und Mann vorgesehen, d. h. die Truppe kommt an diesen fünf Tagen nicht mehr unter Dach und Fach, sondern befindet sich entweder im Gefecht oder aber auf dem Fahrzeug, um je nach den Erfordernissen der Lage an einem anderen Brennpunkt eingesetzt zu werden. Die Uebungen der letzten Tage werden sich im Raum zwischen Freiberg in Sachsen, Tippoldiswalde, Pirna, Hohenstein und Königstein abspielen. Am 18. September trifft das Regiment wieder in Berlin ein.

Eine Woche darauf, am 25. September, veranstaltet es zum Abschluß seiner im Laufe dieses Spätsommers nach wehrsportlichen Grundsätzen durchgeführten Wettkämpfe einen sogenannten Abschlußtag, der in Reinickendorf abgehalten wird und zu dem sämtliche Offiziere, Unterführer und Männer sowie die Arbeiter und Angestellten die beim Regiment waren bzw. sind, mit ihren Angehörigen zusammenkommen sollen.

*

Heute zwischen 5 und 7 Uhr wird das Regiment aus der Richtung Camburg—Dornburg in Jena eintreffen. Die Jenaer Bevölkerung und vor allem die begeisterte Jenaer Jugend werden der einrückenden Truppe einen warmherzigen Empfang bereiten. Es sei wieder darauf hingewiesen, daß bei der schnellen Beweglichkeit der motorisierten Truppe die Straßen unbedingt freizuhalten sind und daß alle Anweisungen der Ordnungsbeamten sofort und unbedingt zu befolgen sind. Die Quartierwirte können also von etwa 6 Uhr an mit dem Eintreffen ihrer Soldatengäste rechnen.

Voraussichtlich rückt das Regiment im geschlossenen Verbande ein, so daß ein prächtiges militärisches Schauspiel zu erwarten ist.

Regiment General Göring Today in Jena!

The Regiment General Göring will leave Berlin in the early hours of September 5 to begin, fully motorized, with 200 vehicles, a large fall maneuver in Thuringia, Bavaria, and Saxony. Taking part in the exercises under the command of Lt. Colonel Jakoby, commander of the regiment, and in addition to the men stationed in Berlin, will be all other formations, I.e. the regimental staff and its attached formations as well as three battalions. These maneuvers, which will average a daily speed of 200 kilometers are regular exercises in regimental formation. During the first days a field march under assumed air attack will take place under the presumption that the Regiment had remained in Berlin as a reserve to the High Command, and was alarmed in the afternoon of 4 September, then ordered to be available for deployment by an army group at the line at the Main river and to reach Jena on 5 september. The exercises of the following days are tailored to determine the endurance of vehicles, materiel and men. The expected development is that the Regiment will be moved closer to the battle front on the Main on 6 September and will be deployed in the fighting in the area around Wurzburg in the night of 8 September..

After the days of rest in Wurzburg the Regiment will take part in the opening of the party congress on 11 September at 11 a.m. at a parade at the Luitpold Hall and before its chief, General Göring, and the Fuhrer. While the party congress is still in session the Regiment will mount its vehicles and move to Hof in Bavaria the same day for further maneuvers. The five days from 11 to 15 September inclusive will put the highest possible demand on men and equipment, i.e. the unit will not be under a roof during any of that time but will be in battle or in their vehicles to be deployed as battle developments require. The exercises of the remaining days will take place in the area between Freiberg in Saxony, Dippoldiswalde, Pirna, Hohenstein and Konigstein. The Regiment will be back in Berlin on 18 September.

One week later, on 25 September, it will host, at the end of its military sport games during late summer, a so-called 'final day' in Reinickendorf to which all officers, ranks and men as well as workers and employees of the Regiment and their families are invited.

Today, between 5 and 7 p.m. the Regiment will arrive in Jena from the direction of Camburg-Dornburg. The people of Jena, in particular its enthusiastic youth will offer the troops a warm welcome. It must again be mentioned that the streets, because of the speed of the motorized unit, must be kept clear under all circumstances and that the directives of the police must be followed fully and without delay. The hosts of the quarters can expect the arrival of their soldier guests from 6 p.m. on.

It is expected that the Regiment will move in as a unit so that a splendid military spectacle can be enjoyed.

The RGG, a Special Luftwaffe Unit

In September 1935, the integration of the State Police Group GG into the air force with a reorganization to the RGG was ordered. The decree of the Reichminister for Aviation and Commander of the Air Force, issued as 'top secret', gave detailed orders on the integration and established the future responsibilities of the RGG. Part of the latter was the guarding of the Headquarters of the Commander of the Luftwaffe in war and peace, including defense against low level enemy attacks in case of war; further the guarding of the official residence of the Commander-in-Chief of the Luftwaffe in Berlin and his offices in Karinhall as well as the responsibility for ceremonial duties of the air force in the capital of the Reich.

These tasks determined the nature and extent of the formation of the RGG. The reorganization occurred in October 1935 at the army training ground of Altengrabow near Magdeburg. The fully motorized RGG was planned to have a strength of 108 officers and 2,935 non-commissioned officers and men as well as 98 horses: the Regiment staff with music corps, communications platoon and cavalry platoon, the 13th Motorcycle Company with an additional light tank platoon, the 14th Engineer Company, further the I. and II. Rifle Batallions with a communications platoon, an engineers platoon, three rifle companies and a machine-gun company each, finally a light anti-aircraft (Flak) unit with a staff battery, two anti-aircraft batteries with twelve Flak 2cm each, and an anti-aircraft battery with six Flak 3.7cm.

The required guard duties, as soon became apparent, burdened the units in an unacceptable manner. This fact led soon to the establishment of a Special Guard Company, the 15th Company, which was followed, within the year, by the establishment of another one, the 16th Guard Company.

In accordance with the diversity of the services within the RGG there were a number of ranks: riflemen (Jager), gunner (Kanonier), engineer (Pionier), motorcycle-rifleman (Kradschutze), tank gunner (Panzerschutze), wireless operator (Funker), cavalryman (Reiter), corporal, sergeant, first sergeant, and finally, sergeant-major.

The change-over from the green uniform of the State Police to the blue of the Luftwaffe could only take place in April 1936. In the meantime, however, to indicate that the RGG was a part of the air force, the air force emblem was worn on the cap and jacket, and the oak leaf garland with wings was added to the cap.

As the only unit in the Luftwaffe the soldiers of the RGG wore on their Luftwaffe uniforms white collar patches, "Weissen Spiegel", with particular pride in the conviction that the RGG was the premier regiment of the Luftwaffe. Instead of the green sleeve band of the State Police, a blue sleeve band with the inscription "General Göring" was now worn, however, on the right foresleeve.

In April 1936 the required manpower level of the RGG was achieved through acceptance of recruits who had volunteered to serve in the RGG. Since the RGG in the future also was to be constituted only of volunteers it was entitled to advertise for future members. For this purpose the RGG established a recruiting office which, together with the selection committee and the music corps, conducted an extremely successful recruiting campaign throughout the German Reich. Thus, all German dialects were soon to be heard in the barracks of the RGG.

The number of volunteers for duty in the RGG greatly surpassed demand and allowed a strict selection process based on physical, mental and moral guidelines.

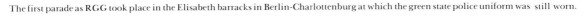

The first parade as RGG took place in the Elisabeth barracks in Berlin-Charlottenburg at which the green state police uniform was still worn.

On the sign: German boy, if you want to wear the white patches of the air force and be our comrade, volunteer right away and join us.

Only volunteers served in the RGG. Its own recruiting office ensured required numbers, later also for the subsequent HG units. Numbers of volunteers were high, the recruitment criteria were stringent as was the selection.

Those who had passed all tests and the selection process were glad to receive the acceptance form.

On the form: Acceptance form for Friedrich-Wilhelm Schade, born 18.6.18, apprentice at Spitzenmuhl/b. Weihmunster. You are herewith accepted as a volunteer for active duty from the autumn 1939 with the Rgt. General Göring at Berlin-Reinickendorf. You are required to register in person or in writing with the responsible Defense Command II Frankfurt/Main within one week by submitting this form. You will be called up later by the Defense Command through a draft order. These directives must be followed explicitly. this form remains valid until autumn 1939.

Regiment General Göring

Freiwilligen-Annahmeliste Nr.

Berlin-Reinickendorf, **21. Jan. 1939**   193
Spandauer Weg 42

Annahmeschein

für _Friedrich-Wilhelm Schade_ geb. am _18.VI.18_
(Vor und Familienname)

Luftprakti'Kant zu _Spitzenmühl/b. Weihmünster_
(Beruf)   (Wohnort, Wohnung)

Sie werden hierdurch als Freiwilliger zum aktiven Wehrdienst ab **Herbst 1939**
(Einstellungstag)

beim **Rgt. General Göring** in **Berlin-Reinickendorf** angenommen.

Sie haben sich innerhalb von 1 Woche unter Vorlage dieses Annahmescheines bei dem für Sie zuständigen

Wehrbezirtskommando _II Frankfurt/Main_ persönlich oder schriftlich zu melden.

Ihre Einberufung erfolgt später vom Wehrbezirtskommando durch Gestellungsbefehl.

Umstehende Anordnungen sind genau zu beachten.

Dieser Annahmeschein behält seine Gültigkeit bis zum **Herbst 1939**
(Einstellungstag wie oben)

(Unterschrift des Kommandeurs des Annahmetruppenteils)

The RGG, Origin of the German Parachute Troop

At the same time the State Police Group GG was incorporated into the Luftwaffe. It was ordered that a battalion was to be trained as a parachute battalion. It was meant to be the origin of a future German parachute force.

When this order was conveyed to the Regiment, together with a request for volunteers, the numbers volunteering were far greater than the demand. The future 'jumpers' were moved to the I. Jager Bataillon, which initially kept this designation for reasons of secrecy, and to the 14. Pionier-Kompanie. Strenuous training followed. Before a Ju 52 could be entered for the first jump, a comprehensive ground training program had to be endured. The practice jumps were done in the Gosener Heide between Berlin and Konigs Wusterhausen, near the Rudesdorfer Kalkbergen, on the field and pastures between Velten and Botzow as well as other suitable locations. The aircraft departed the airfield Schonwalde near Berlin where the jumpers had been moved by vehicle.

After six different jumps, the young paratrooper received the parachute jumper certificate which was coupled with the awarding of the parachute jumper medal. Since it had been earned through hard work it was worn with justified pride. Different from today, at that time the aircraft was left by jumping head forward which demanded an extreme amount of self control. Only real men with fearless hearts could satisfy the high demands. This was one of the reasons which led, together with the outstanding training, to the feeling of elite within the parachute troops.

With the renewed changes in the RGG at the end of 1937 the need for secrecy disappeared and the official designation could be used. The I. Jager-Batallion became IV. Fallschirm-Schutzen-Batallion (paratroop battalion), unofficially also called Fallschirm-Jager-Batallion. It remained with the RGG until 31 March 1938. At that time it was renamed I./Fallschirm-Jager-Regiment 1 and received the yellow patches of the Fliegertruppe (aviation troops) as well as a green sleeve band inscribed "Fallschirm-Jager-Regiment 1". It was moved to Stendal in the middle of 1938.

From this first parachute battalion of the RGG, and the army in general, prominent paratroopers originated who put their stamp and direction on the German parachute troops. To name only a few, the commander of this battalion, Major Brauer, and other members such as Walter, Reinberger, Kroh, Vogel, Schulz, Groschle, Gericke, Koch, Paul, Noster, and Dunz.

Many of the sergeants of the battalion were later promoted to officer, a great number of them for courage in the face of the enemy, and quite a few of them achieved commands of their own. Of these first paratroops, many achieved commands of their own. Many also achieved high awards for courage during the war. A large number of them, however, lost their lives to the enemy.

Leaving the Ju 52 head-first. This required an unusual measure of self control. Only men with fearless hearts were able to withstand the physical and mental demands. The great successes of the German paratroops in the Second World War, were largely due to the stringent selection standards and the high demands which achieved the self confidence of this elite troop which had its beginning in the RGG.

Text of Letter: From the Minister of the Reich for Aviation and Supreme Commander of the Luftwaffe. (dated) Berlin, 29 January 1936. ''Secret'' To the Luftkreiskommando II (air defense command) Berlin.

In preparation of parachute training for the Regiment ''General Göring'' it is ordered: 15 officers, sergeants and corporals, based on volunteers, are to be made available and trained. Acceptable are men weighing less than 85 kg. (including clothing) who are physically fit and meet aviation medical requirements. Start of training: Planned for 1.4.1936. Duration: 8 weeks, 4 of which as parachute inspector with the inspection unit, followed by 4 weeks of field training in parachute jumping from aircraft. Planned airport is Neubrandenburg. A Ju 52 will be provided by R.L.M. (L.C.) (R.L.M. 'Reichs Luftfahrts Ministerium). Qualified instructors will be provided by R.L.M. (L.A.III). On 15.3.36 the L.K.K. (Luftkreiskommando) will report 1. Ranks and names of the volunteers, 2. The completion of the aviation medical examinations, 3. The usability of the planned airfield. If necessary, other proposals are to be submitted. The extra costs to the command are to be requested from Kap.A2 Tit.34 subsection 4b, they will be separately transferred. For the duration of the four week field training in parachute jumping the special aviation allowance is authorized.

Signed, Milch

The RGG was the source of the first German parachute troop. The first soldiers were selected for parachute jump training at the beginning of 1936. Instead of the initially planned fifteen soldiers, the first course trained more than thirty jumpers. This course was not run at the airfield of Neubran denburg but at Stendal airfield, seat of the future school of parachute jumping.

The training of the first parachute jumpers took place in Stendal during May/June 1936. The first course included men who are part of the history of the German parachute troops, among others: chief instructor Captain Immans (in white jacket), to his left battalion commander Major Brauer, further the Captains Reinberger and Kuhno, the First Lieutenants Dunz, Kroh and Rag, Head Physician Dr. Greiling, the Lieutenants Moll and Schelske. In the front row, second from the left, the specialist in parajumping, from the aviation authority, Diete, further the Sergeants Loos, Megow, Minkwitz and Zinke and Sub-Lieutenant Kiess, the Corporals Adryan, Dochow, Hasse, Helmbold, Kroll, Lienek, Schulze, Volkmann and Wirth, Lamce-corporals Braunshirn, Patsch and Rufleth and the Privates Bauer and Hagedorn.

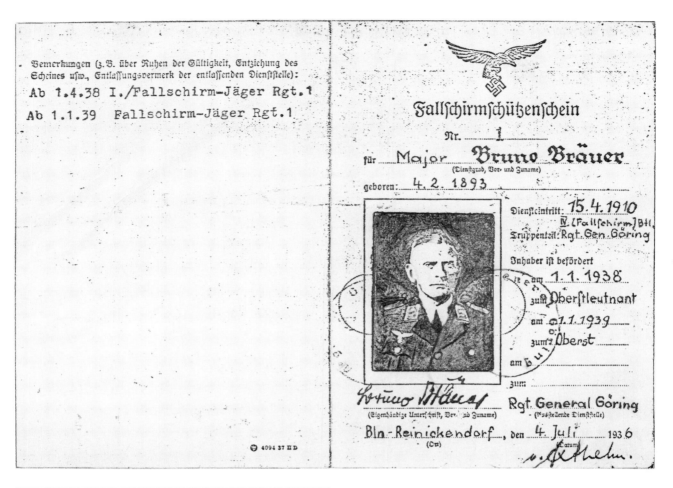

Ab 1.4.38 I./Fallschirm-Jäger Rgt. 1

Ab 1.1.39 Fallschirm-Jäger Rgt. 1

4094 37 II D

Fallschirmschützenschein

Nr. 1

für Major **Bruno Bräuer**
(Dienstgrad, Vor- und Zuname)

geboren: 4.2.1893

Diensteintritt: 15.4.1910

IV. (Fallschirm) Btl.
Truppenteil: Rgt. Gen. Göring

Inhaber ist befördert

am 1.1.1938

zum Oberstleutnant

am 1.1.1939

zum Oberst

am

zum

Rgt. General Göring

(Eigenhändige Unterschrift, Vor- und Zuname)   (Ausstellende Dienststelle)

Bln.-Reinickendorf, den 4. Juli 1936
(Ort)   (Datum)

A Sample of the Parachute Jumper's Certificate.

The first jump by parachute by a German soldier took place on May 11, 1936 during the first jumping course. It was executed by Major Brauer who jumped off the wing of a Klemm Kl 25. After completion of the course he received certificate no. 1 and was, from then on, called by his men "Paratrooper No.1".

With receipt of the jumper's certificate came the awarding of the paratrooper's badge which, as it had been earned with hard work, was worn with justified pride.

The Barracks of the RGG in Berlin

Life in the barracks did not initially change for the detachments and units after the move to the Luftwaffe. Remaining in the Elisabeth barracks in Berlin-Charlottenburg were the regimental staff and the I. Jager-Batallion, in the Moritz barracks in Berlin-Spandau the II. Jager-Batallion and in the aviators' barracks in Berlin-Reinickendorf the III. Jager-Batallion which had now become III. Flak-Abteilung (anti-aircraft detachment).

In order to achieve the required quartering of the total regiment a new barracks was built in Berlin-Reinickendorf, at the edge of the Tegel rifle range and taking in the aviators' barracks dating from the time of the Kaiser, to be ready for occupancy in September 1937. it became, in its efficient but not expensive manner, the example for later barracks of the air force. Truly a home to the soldiers, it contained more than 120 buildings with gymnasiums, an indoor swimming pool, an outdoor pool with a ten meter diving platform, an arena, a postal station and many other amenities.

At the edge of the former Tegel rifle range, in the Berlin district of Reinickendorf-West, a new barracks was built in 1936/37 to completely house the RGG. Celebrating the erecting of its highest spot, all of the Regiment fell in at the building site. In the evening, soldiers and construction workers showed their unity in the Deutschlandhalle (German hall) enjoying free beer, sausages and potato salad.

The entrance to the barracks of the RGG which was then the most beautiful home to soldiers in Germany. In the back the building of the regimental staff.

Building of the regimental staff. Above the door to the main entrance was carved the motto of the **RGG** on a marble slate:

"Courage, obedience, honor, comradeship are the basis of soldiership".

Living quarters of the artillery.

The barracks of the RGG, with its simple styling typical for the area, fit perfectly into the terrain of the Mark. The buildings were of architectural beauty without expensive or even luxurious formation. Within the barracks grounds, in addition to an indoor pool, there was an outdoor pool with a ten meter diving platform, especially fitting for paratroopers! Today, a pool within a barracks is nothing unusual, then it was sensational.

Entrance to the 'Schweine-Kantine' (pig-canteen) in the supplies building of the I. Abteilung.

The ring road, the Barbararing; left, a building of the artillery, in the background one of the supply buildings.

18

Each kitchen could feed more than 1,000 soldiers and was the responsibility of a kitchen sergeant. In addition to civil workers, soldiers were employed as part of their training as future cooks in the field.

Each detachment had a state-of-the-art workshop to maintain and repair all weapons, from the pistol to the 8.8 cm Flak (anti-aircraft gun), by civilian and military specialists. They were under the supervision of the 'Waffenrevisor' (head of the workshop).

In the modern vehicle repair shops civilian workers maintained the many vehicles of the regiment. Soldiers received here their training for the maintenance of the vehicles entrusted to them.

Entrance to the Velten camp with guard buildings.

In summer 1939, the IV. Abteilung was removed from the RGG and, under the name 'Leichte Flak-Abteilung 79' (light anti-aircraft detachment), integrated into an army division. The regimental commander was able to wear the white collar patches and the sleeve band, which had been their wish all along.

The Years of Peace

After the reorganization, a period of stringent training began in the RGG. It was necessary now to train the men on new weapons, to strengthen the spirit of the newly formed units and then to train the newly arrived recruits. After the basic training, training in the field soon began, now also at the training grounds of the regiment between Hohenschopping and Velten north of Berlin which bordered the Velten camp. Later followed exercises of larger units close by and further afield during which the soldiers, who had come from all the German provinces, came to know the beauty of the Mark Brandenburg.

In the capital of the Reich, the soldiers with the white patch and blue sleeve band were soon well known, either while marching with or without music, on horse or vehicle, even after duty in dress uniform. During the summer with white cap and gloves, the more vain ones even in white pants.

During time off, the lively metropolitan Berlin at that time offered many diversions. Potsdam too, and other locations of historical significance to Germany and Prussia were popular places to visit and such visits were encouraged by the regiment.

The games of the XI Olympiad held in Berlin in August 1936 were a special experience for many member of the RGG.

On special days of remembrance such as the annual 'Day of the Air Force' (1 March to commemorate the 'de-camouflaging' of the air-force in 1935), on the day commemorating the re-introduction of the draft on March 16, 1935, and on the 'Day of the Armed Forces' celebrated once a year collectively by all units in all of the Reich, marches, parades and exhibits took place. On special occasions the 'Great Tatoo' (Grosser Zapfenstreich) was performed. The RGG took place in all these activities, which were greatly enjoyed by the population, not only in the barracks and drill grounds, but also, and in particular, in public surroundings.

Naturally, the music corps and its musicians (because of their dashing appearance) were especially notable. The music corps, with forty-six sergeants and fourteen enlisted men, was the largest in numbers in the Luftwaffe and was led by the head conductor, Stabsmusikmeister Paul Haase. He introduced, in the search for new forms of music for the air force (softer sounds), saxophones into his music corps and thus, for the first time, into any German music corps. Saxophones in a military band-then a sensation, today an accepted practice. Stabsmusikmeister Haase became professor of military music at the academy of music in 1942. His successor was Obermusikmeister Hans Friess.

The RGG provided daily the guards for the staff headquarters including the domicile of the supreme commander of the Luftwaffe. This was located, connected to the newly built ministry of aviation of the Reich, in a park area behind the former parliament of Prussia and former 'upper house' of Prussia which had become, in 1935, the 'House of the Aviators'. The guard was moved by vehicle, together with the music corps, to the Hofjagwerallee and then marched, accompanied by rousing music, to the Prinz-Albrecht-Strasse where the entrance to the headquarters was located, for the changing of the guard. Every Wednesday there was an especially ceremonial changing of the guard. It began at the police barracks in the Universitatsstrasse, led to the Strasse Unter den Linden, past the memorial of Frederick the Great, to the 'Reichsehrenmal' (memorial of honor of the Reich). Every Thursday the Leibstandarte performed the ceremonial march, on the other days it was the Guard Regiment of the army, which later became the infantry regiment "Grossdeutschland". Once every year, on 'Skagerrak—Day', May 31, the navy performed at the memorial.

Göring had built a retreat headquarters in the Schorfheide, some fifty kilometers north of Berlin, which was named after his first, deceased wife Karin, "Karinhall". The 'Waldhof' which comprised the offices, meeting, dining, and living quarters was located just off the highway Berlin-Prenzlau, between the lakes Grossen Dolln and Wucker-See. The responsibility for guarding the compound was also that of the RGG. To achieve this during peace time there was a platoon of the guard company of forty to eighty men (strength depending on security requirements) detailed on a continuing basis. Change of personnel occurred on a daily rotation, later, after required accommodations had been built, after longer periods.

In March 1938 the RGG took part, without the guard battalion, in the march into Austria and remained, until the middle of April, in Wiener (Vienna) Neustadt.

During the Sudeten crisis in late summer of the same year the RGG deployed to positions in an area west of Potsdam to guard the headquarters of the supreme commander of the Luftwaffe which had been moved to the air war school Wildpark-Werder. Portions of the guard battalion of the RGG and the Luftlande (airborne) Battalion Sydow, which had been established by the RGG as an experimental unit, saw action in the Sudetenland in the framework of the 7. Flieger-Division near Jagerndorf-Freudenthal.

The RGG also took part in the occupation of Czechoslovakia with the march into Prague. The IV. Flak-Abteilung, however, was seconded to the 2. Panzer-Division to secure its moves into action and marched with it through Grafenwohr-Weiden and the Bohemian forest (Bohmerwald) to the area of Pilsen. There, the Abteilung was charged with guarding the Skoda works.

The peacetime units of the RGG organized later (Guard Battalion, III. Searchlight Unit and IV. Anti-Aircraft Unit) were given a flag of the same pattern as the police flag when they were organized.

On April 21, 1936, the memorial day on which, eighteen years before, the fighter pilot Manfred Frhr, von Richthofen had been killed, the II. Jager-Batallion and the III. Flak-Abteilung each received a flag in the same manner as the I. Jager-Batallion already had, namely the police flag which had once been awarded the L.P.G. Wecke z.b.V., their commanders, adjutants and other officers.

After the mobilization the flags of the RGG, together with those of the Reich. At the end of the war that were taken by Soviet troops as war booty to Moscow.

The Regiment in formation at the sporting grounds for the swearing in ceremony. The ceremony takes place in front of the weapons which the recruits will learn to use, in a spirit which reflects the seriousness of the occasion. One recruit from each detachment has been chosen to reinforce his oath by touching the flag.

"3. Gruppe-Richt Euch" (3rd group-right dress) sergeant Nuske has ordered. Training follows the old and proven principle 'demonstrate, imitate, practice' and was the charge of the group leader with a lance corporal as his assistant.

The cavalry platoon, later cavalry squadron, was unique and was the only mounted unit in the Luftwaffe. The training was varied and all-encompassing. Here a cavalry troop in difficult terrain.

Guard duty is also a part of it. Before the change of guard, uniforms and weapons are checked.

Training on the listening device, the circular-funnel direction listening device.

Soon there was more training in the field.

A break in the march.

Maintenance of weapons was high on the list of priorities while on maneuver, cleanliness and completeness of the weapons was assured through frequent checks.

Troops on maneuver often enjoyed concerts by the music corps. In the picture, horn player First Sergeant Schulz with his accompaniment whose exact movements caused great admiration with the spectators. He, as his successor First Sergeant Bogenschneider, was as the saying went then, "an arrow-straight figure with spit-shined boots". Even today the legend survives of First Sergeant Schulz having answered the then Generaloberst Göring, when asked to stand at ease after reporting, "Sir, standing at attention is the most comfortable to me".

Excercises of a few days with bivouac were very popular.

The time spent every year at the various training grounds and firing ranges was a welcome diversion. These are the barracks at the Deep firing range in Pomerania.

The searchlight detachment practiced cooperation with the heavy anti-aircraft gun detachment.

Using live ammunition, the gunners were able to determine the success of their training by the number of hits. Particularly impressive was the first live firing of the "Achtacht" (8.8 cm gun).

On March 1, 1938, the 'Day of the Air Force', a parade of all branches of the air force took place in Berlin. Leading it was the 10. (Wach-) Kompanie (guard company) of the RGG led by Hauptmann Kluge. Front right is sergeant Scheid who was awarded the Knight's Cross (Ritterkreuz) in Africa. The flag on the left is that of the RGG. The officer on the left is Oberleutnant (first lieutenant) Brandenburg.

The supreme commander of the Luftwaffe inspects the guard battalion of his regiment, lined up in front of the ministry of aviation of the Reich, in April 1938. Behind him: the Generals Kesselring, Stumpff and Weisse, Oberst (Colonel) Bodenschatz (covered) and the commander of the RGG, Oberleutnant (Lt. Colonel) von Axthelm.

To commemorate the day in 1935 on which the newly created Luftwaffe was unveiled the "Day of the Luftwaffe" was celebrated annually on March 1. In the photo, the RGG march past on the 'Day of the Air Force' 1939 at the ministry of aviation of the Reich in the Wilhelmstreet, before the Supreme Commander of the Luftwaffe, Göring, since April 21, 1938 Generalfeldmarschall. Music corps and band played the parade march of the RGG, the march "Der Jager aus Kurpfalz" (The hunter from Kurpfalz) by Gottfried Rode.

Among the responsibilities of the RGG was the guarding of the official residence of the minister of aviation of the Reich and supreme commander of the Luftwaffe as well as that of his headquarters located in the same building. The villa was situated in the park area between Leipziger Strasse and Prinz-Albert-Strasse and had been for years the official residence of the Ministerprasident (prime minister) of Prussia and had been used by Göring when he held that office.

Every Wednesday the RGG provided the guard for the memorial of honor of the Reich. The mounted chief of the Kompanie commanded the spectacle which was always much enjoyed by the Berlin populace. In the photo Hauptmann (captain) Funck with his Kompanie marching through the Universitatsstrasse.

The RGG parading past the memorial of honor of the Reich. In the lead Oberfeldwebel (first Sergeant) Bogenschneider with his forty band members. Behind is the music corps, sixty strong, led by Stabmusikmeister Haase. The photo was taken on August 25 when the Reichsverweser (governor) of the Kingdom of Hungary, Admiral von Horthy, was in Berlin on a state visit. The monument to Frederick-the-Great was returned to its old spot Under den Linden, as then in 1938, only in 1981.

Every year in April the RGG took part in the big parade of the Wehrmacht (armed forces) at the East-West-Axis Road in front of the Technical University of Berlin. Here the march past of a Zug (platoon) with listening devices of the 11. Scheinwerfer (search light) Batterie of Hauptmann Meyer during the last peace time parade.

The recruits from all of the German Reich, mostly in groups, came to know their garrison city of Berlin and the outstanding points of interest of the Mark Brandenburg. Here, Oberleutnant (first lieutenant) Brandenburg leads a group through Sansouci (castle built by Frederick-the-Great).

On March 11, 1938 the RGG was first sent into action towards an initially unknown destination. En route it was made aware of the destination: Austria. In three days of marching it reached Bruck on the river Laitha, not far from the Hungarian border. A week later it moved to Vienna and Bad Fischbach. In the photo, the RGG is ready to move out of its barracks. The heavy Flak-Abteilung had received, shortly before, as replacement for the three-axle Henschel trucks, brand new tractors on tracks.

The 8. (Kradschutzen) Kompanie of the RGG, "the fastest of the fast", under the leadership of Major Weber, raced ahead and reached, as the first unit of the RGG, its destination, Austria.

In Wiener (Vienna) Neustadt, opposite the venerable military academy Maria-Theresia, a joint parade with Austrian forces took place.

A few days of rest allowed the opportunity to get to know the beauties of the 'Ostmark', as that part of the country was then called. A short break during an excursion by members of the I. Flak-Abteilung. Among others: Schroder, Nuske, Magnor, Grosse, Schmudlach, Wagner and Otte.

During the action in Bohmen (Bohemia) and Mahren (Moravia) the field post/mail service was used for the first time.

"Finally, cheerfulness (and not as a worthless possession) must be expected from the soldier. Those in charge have the duty to foster the 'joy of the business'. With the desire for enjoyment and sense of humor, everyone engaged in the soldier's life will be party to unlimited hours of satisfaction". (General-oberst (4 star general) Hoepner).

The evening formal dress, 'Fliegerfrack' (fliers' tux) for short, was obligatory for all officers of the RGG.

The 'watering hole' in the sergeants' quarters where quite a few promotions were celebrated.

The 'small' dining room in the officers' club of the RGG.

At the mobilization, all of the RGG was inoculated. The regimental physician, Stabsatz Dr. Siebert, gives the shot to his commander, Oberst (Colonel) con Axthelm.

While the gunners fortify their position and dig ditches, according to the motto "Sweat saves blood", an observer watches the airspace above.

The Development of the RGG
to the Beginning of 1940

After moving to the new barracks in Berlin-Reinickendorf an all-encompassing reorganization of the RGG followed. The emphasis, until then primarily on infantry tasks, as changed to that of the air defense. Apparently, this change was due to the influence of the new Regimentskommandeur (commander of the regiment) Oberstleutnant von Axthelm, who had taken over the RGG in August 1936 from Oberstleutnant Jakoby.

In October 1937 the following were created:

The I. schwere (heavy) Flak-Abteilung (Hauptmann Hullmann) from the II. Jager-Bataillon.

The II. leichte (light) Flak-Abteilung (Major Conrad) through renaming of the III. Flak-Abteilung.

The III. Wach-(watch) Bataillon (Major Sydow) through combining of the two Wach-Kompanien (watch companies), the Kradschutzen-(motorcycle) companies and the mounted platoon.

The Fallschirm-Schutzen-(paratroopers) Bataillon (Major Brauer) through the revealing and renamingof the 14. Fallschirm-Pionier-Kompanie.

The next significant change in the RGG followed a year later. On November 1, 1938 the anti-aircraft capacity of the RGG emerged: I. schwere Flak-Abteilung (Hauptmann Hullmann), II. leichte Flak-Abteilung (Major Rudel), III. Scheinwerfer-Abteilung (Major von Oppeln-Bronikowski), IV. leichte Flak-Abteilung (Oberstleutnant von Hippel), Wach-Batallion (Major Weber).

At the end of 1939 the previously mentioned Luftlande-(airborne) Batallion Sydow was removed from the RGG and relocated, as III./Fallschirm-Jager-Regiment, to Gardelegen.

In August 1939 the RGG went into its war formation. To the peace time troops were added: the 14. schwere Eisenbahn-Flak-Batterie (heavy railroad anti-aircraft guns) (10.5 cm, Lt. Arnold); the Reserve-(reservist) Scheinwerfer-Abteilung, and the Reserve-Abteilung (Major von Ludwig).

A few other minor changes need to be mentioned: the reinforcement of the Reiter-Schwadron (squadron) in the summer of 1938 and the change of this squadron to a (new) Kradschutzen-(motorcycle) Kompanie in March 1940. There remained in existence a Reiter-Zug (mounted platoon).

The anti-aircraft forces of the RGG were deployed during the Polish campaign to secure the capital of the Reich, Berlin, the other specialties were used in accordance with situational requirements. At the end of 1939 the regimental staff of the RGG was given the code designation Flak-Regiment 103. The regimental staff, the I. and IV. Abteilung was integrated into the newly created II. Flakkorps, the III. Abteilung was integrated into the newly created I. Flakkorps, and these parts of the RGG were moved to the Westwall (western defensive line), to the areas of Aachen and Trier respectively.

The Wach-Kompanie of Hauptmann Kluge was moved to the Warsaw area, partly by air, in September 1939 to guard airports and installations of the Polish aviation industry there.

Other parts of the Wach-Batallion reached their destination by vehicle. Here, entry into Warsaw.

During the first days of the war, the 2cm. guns (Sf self-propelled) of the 8. Batterie, hauptmann Seewald, deployed in Berlin along the east-west axis at the Siegersaule (victory column).

The 8, Batterie was moved, still in September 1939, to East Prussia to guard the headquarters of the Oberbefehlshaber (supreme commander) of the Luftwaffe which was located, during the Polish campaign, in the Reichs-jagerhof (Reich hunting lodge) in the Rominter Heide. The journey there was via Marienburg and passed by the most important castle, the Deutscher Ritterorden (German order of knights).

During the winter 1939/40 the heavy Flak-Batterie needed to open fire only rarely, the enemy launched only a few air attacks.

The headquarters of the Oberbefehlshaber of the Lufwaffe in the Reichs-jagerhof in the Rominter Heide. In the photo: right, Oberst Conrad who became Kom-mandeur of the RGG during the French campaign.

The first Christmas of the war in a warm bunker. During the Christmas broadcast on radio the promise was made: 'You will be back home by next Christmas', it was not kept. Five more war Christmases were to follow.

Timely, before the start of the particularly hard winter of 1939/40, the gunners had prepared their positions for the cold and moved into warm underground bunkers.

The men at the front knew their families to be in good care at home in Berlin. In the hospital of the Reinickendorf barracks, wives and children received any required medical help.

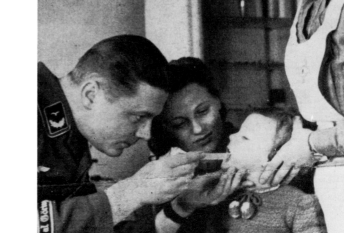

The Detachment Kluge in the Norwegian Campaign

When the war widened to northern Europe on April 9, 1940 and German troops occupied Denmark and Norway (Unternehem (operation) Weserubung) a part of the RGG participated. This part, Detachment Kluge, led by Hauptmann Kluge, was formed from the Wach-Batallion/RGG with a Schutzen-Kompanie and the newly established Krad-schutzen-Kompanie (with Panzerspah Zug-armored scout platoon), as well as the 8./RGG (2 cm. Flak-Batterie, self propelled). In a surprise action the airport and wireless station of the Danish port city of Esbjerg were occupied and later the north coast of Jutland (Jammerbucht) was secured. In the middle of April the Detachment Kluge was moved by sea to Oslo and fought successfully, together with the army (Heer), to achieve the land link to Drontheim. Subsequently, it pushed, with Gebirgsjagers, under the most difficult weather conditions and treacherous terrain, across the arctic circle and finally reached Bodo, to relieve the Gebirgsjagers and Fallschirmjagers, encircled at Narvik. In the middle of May the Detachment Kluge, after achieving its objectives, returned by sea to Berlin.

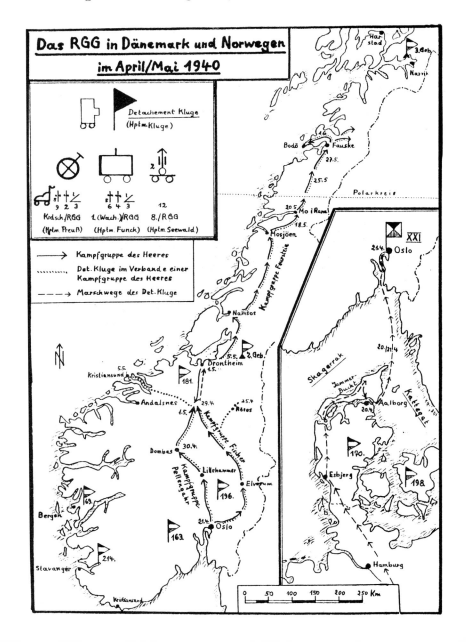

Through such deep valleys, snow covered and icy still in April, and offering all advantages of such terrain to the defenders, the Detachment Kluge, together with troops of the Gebirgs-Division of General Dietl, continued its advance from Drontheim across the Arctic Circle. To detour the countless barricades of trees and rock across roads was mostly impossible. The barricades had to be removed, often under enemy fire, before the advance could continue.

In April 1940 the Detachment Kluge (without the 8./RGG) was transferred to Norway. In the photo, the Kradschutzen-Kompanie is loading its vehicles onto the freighter "Campina". Only the drivers remain on board. The other men are ferried on the torpedo boats "Falke" (falcon) and "Jaguar" to Norway.

The Polar circle has been reached. Seated in front of the marker is Lt. Gerhard, one of the Zugfuhrers (platoon leader) of the Kradschutzen-Kompanie. But soon they will go even further north.

After the completed action the Detachment returned to Germany. In the port of Oslo, from where the steamship "Bahia" departed with the Detachment on June 18, 1940, the hospital-ship "Wilhelm Gustloff", with wounded on board , was ready to cast off.

The RGG in the Western Campaign

On May 10, 1940, the units of the RGG, in readiness in the west, also saw action. In coordinated advances with the Panzerverbanden (Panzer units) of the Heer and the motorized Korps of the 6. and 4. Armee the Flak-Batteries of the RGG were able, for the first time, to demonstrate the tank-destroying force of their guns and to keep their field of action free of enemy aircraft. The crossing of the Maas River, the fighting with the retreating opposition forces in east Belgium, the crossing of the Albert canal, the breakthrough at the Dyle position, the capture of Brussels were the scenes of action for the RGG in the first phase of the western campaign. The battle in Flanders and in Artois with the tank battle of Gembloux and the wrestling for the Mormal woods showed that the RGG was no parade unit, rather a unit trained for battle which, together with the fastest troops of the Heer, always kept the lead. During subsequent fighting at the Somme and Aisne rivers, the RGG again showed its mettle.

After completion of the western campaign on June 26, 1940, the RGG took over the guarding of the Channel coast and later of the airspace over Paris.

In September 1940 the RGG returned by rail to Berlin and was there again deployed to secure the airspace while the Regimentsstab (regimental staff) formed the staff of the Flakgruppe West.

Through Holland and Belgium into France. Pioneers quickly built bridges across the many canals in Holland and enabled the surprising advance of the German troops.

The Batterie-Chef (Battery Chief) briefing the men.

Despite fortifications and the courageous resistance of the Belgians, the crossing of the Albert canal was achieved.

In the wood of Mormal, on May 20, 1940, a dramatic duel between Flak and French tanks ensued when both parties faced each other separated by only a few meters. Among others, and receiving their baptism of fire, the gun "Casar" (Caesar) of the 3./RGG and a 2cm. gun of the 5./RGG took part. This photo was taken during the fighting. On the left a part of the "Casar" is visible.

The victorious crew in front of its "Casar" after the duel: Geschutzfuhrer (lead gunner) Kubaschk, the Kanoniere (gunners) Moller, Radke, Rethmeier, Peyer, Lubberstedt, Diekneite, Braschwitz, Korner, Stell and Schutz.

"Casar's" opposition were tanks of this type: D2 Renault, weight 18 tons, 4 man crew, 4,7 cm gun, 2 machine-guns, 25-40 mm armament, speed 30km/h, introduced into the army in 1936.

A short pause in the march is being used to get a quick bite from the field kitchen and to enjoy "forty winks". The use of the Flak-Abteilungs in the framework of the Panzer-units of the Heer posed extremely high physical demands. Terrain and air defense objectives followed each other in quick succession. Frequently, two or three times per day, positions had to be changed. There was almost no chance to rest.

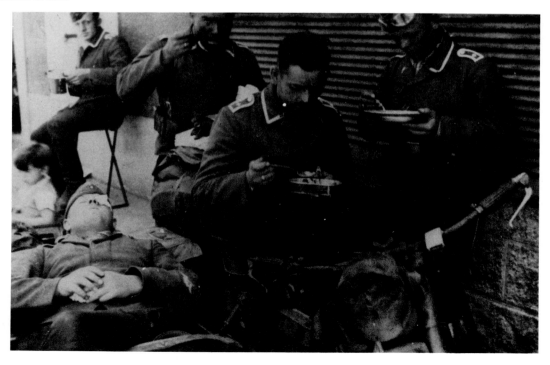

Without break, the columns of the defeated French army, many colonial troops among them, stream backwards past the advance route.

The Dutch and Belgians had capitulated after a few days, the English had gotten away through Dunkirk, and now, after some six weeks of fighting, France was at the end of its strength. On June 21, 1940, in the woods of Compiegne, the armistice was signed. The French delegation arrives for the armistice negotiations. The Ehren-kompanie (honor guard) of the Fuhrer-Begleit-Batallion (Fuhrer's personal battalion) is formed from two platoons of the Heer and one Zug of the RGG under Oberleutnant Dieke which is deployed as Flak-Batterie at the Fuhrerhauptquartier (Fuhrer's headquarters).

"Victory in the West-Armistice" the just-distributed field newspaper proclaimed. The gunners on their half-tracks acknowledge this message with inner excitement. At the beginning of the western campaign the white collar patches had to be removed for reasons of camouflage. For the same reason the RGG was temporarily given the camouflage designation "Flak-Regiment 103".

Paris quickly became the seat of many military headquarters.

After the campaign in the west, the RGG was deployed to guard against aerial attacks, and also at the airports of Villacoublay and Orly which were close to Paris. From here, in the framework of recreation, visits to Paris and Versailles were conducted.

During readiness in the gun position a vigorous 'Skat' (card game) was sometimes played.

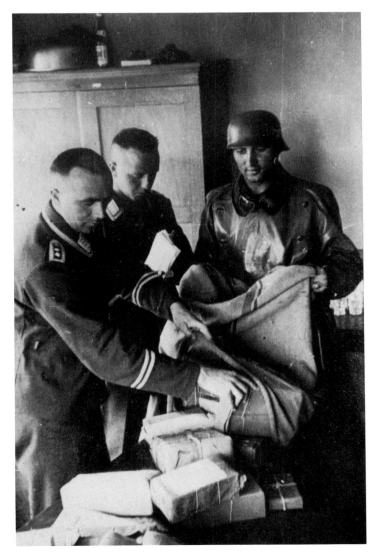

Field mail was always eagerly awaited.

The English flew in mostly at night. There was a lot to be done then.

The visible results of our "night work".

On the channel coast, too, vigilant readiness.

In the autumn of 1940, the air raids on Berlin became more frequent, which required the reinforcement of the air defense forces there. For this, the RGG was tagged and brought back by rail at the end of September.

In Action in Romania

When, on April 6, 1941, and quite unexpectedly, the Balkan states were drawn into the war, the RGG saw action there also. The I. and IV. Flak-Abteilung were pulled from positions around Berlin, and from the Wach-Batallion, a Schutzen-Batallion (commander: Hauptmann Frank) was set up. Under the leadership of the Regimentsstab (regimental staff) these units were sent to the new theater of war. Having arrived in Romania by ground transport, the RGG was placed under the command of the XXXXI. Armeekorps, which belonged to the 12. Armee. It was a reserve to the Armee and did not see action during the Balkan campaign. The RGG took on the defense of the oil fields in the Ploetsi area against air raids. The Schutzen-Batallion used this time for concentrated training in view of future requirements as the parts of the RGG in Romania had been planned to be in action in the eastern campaign (Unternehem (operation) Barbarossa). They were shipped in May 1941 to the readiness position near Zamosc (80 km. southeast of Lemberg) and moved forward, in mid-June, to the Bug river.

In the context of the Balkan campaign the RGG, in April 1941 with the Regimentsstab, the I. and IV. Abteilung, the newly created Schutzen-Batallion, and the Kraftfahrzeug-Werkstatt-Zug (vehicle repair and maintenance platoon), moved through Prague, Vienna, and Budapest to Romania. The Hungarian border was reached quickly.

In a few days Budapest was reached. The planned deployment as reserve to the Armee in the Balkan campaign did not occur. Instead, the RGG is charged with guarding the oil fields near Ploesti.

On days of rest, such as here in Sandomir, vehicles and weapons are inspected.

On dusty roads in Poland towards the readiness area of the Panzergruppe Kleist. The vehicles have a "K" painted on them.

In May the RGG was made ready for the eastern campaign. It moved, in stages, via Krakow-Sandomir to Zamosc, and at the beginning of June, to the Bug river near Sokal. Wherever the RGG moved to, disciplined behavior (as in the photo in Krakow) was made a priority.

We did not want to believe rumors of an impending war against the Soviet Union. Now that we had moved into bivouac positions close to the Bug opposite the city of Sokal, these rumors appeared more and more true. It is June 21, 1941. After dark, the group leaders read to the troops the directives of Hitler and the order for the attack the next day, which would become a day of destiny in German history. It was exactly 129 years to the day that Napoleon had attacked Russia and suffered total defeat. An unbearable tension and oppressive uncertainty burdened each one of us. All knew instinctively that hardships lay ahead.

After crossing the Bug we experienced the endless spaces and Russian roads with unimaginable dust, which would cause us many problems.

The RGG in the Eastern Campaign

On the fateful June 22, 1941, the RGG crossed the Bug near Sokal. It was deployed in the section of the II. Flakkorps and fought here, as during the western campaign, with the Panzergruppe von Kleist, allocated mostly to the XXXXVIII. Armeekorps and almost always to the 11. Panzer-Division. The breakthrough at the Bug, the tank battle of Redziechow, the advance towards Dubno, the encirclement battles of Kiev and Brjansk (and for parts of the RGG), the action near Orel, and the securing of the operations area during the hard winter 1941/42, were the milestones of a hard fight during which the RGG, too, had to take regrettable losses.

In December 1941 the RGG was withdrawn as being "worn out", and returned by rail without its guns and heavy equipment to the Reich. The two Flak-Abteilungs were re-equipped and deployed to Munich for air raid protection. In March 1942 the RGG was reorganized and received the designation "verstarktes Regiment (mot.) HG" (reinforced Regiment (motorized) HG), in short: RHG.

At the end of 1941, in Berlin, a second Schutzen-Batallion/RHG was formed from the Wach-Batallion/RHG and sent to the eastern front where it saw action until April 1942 in the fighting for Juchnow and Anissowo-Gorodischtsche and was almost completely wiped out. Its small remains were moved to France and there the Batallion was re-established.

Because of a major air raid by the Allied air force on the Renault works near Paris at the end of April 1942, the two Flak-Abteilungs of the RHG were moved from Munich to the area Paris-Versailles. They remained there until May 1942.

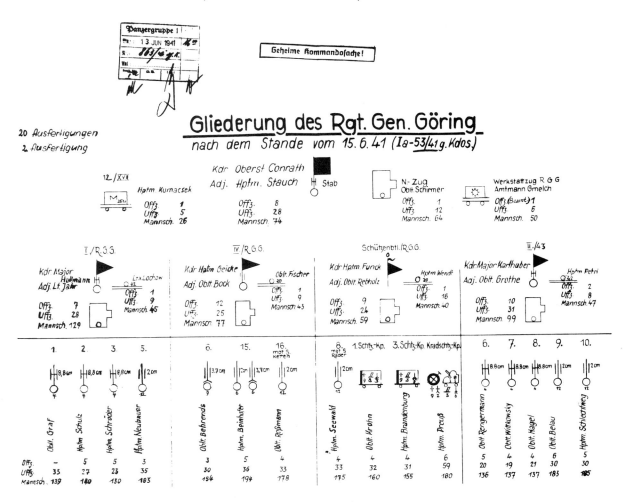

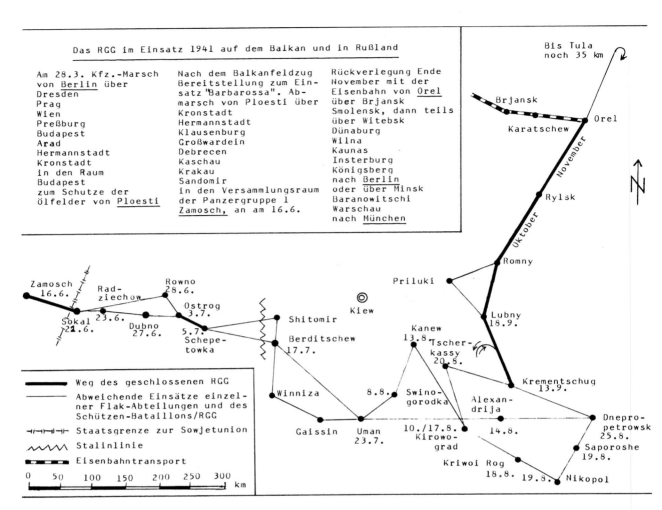

Das RGG im Einsatz 1941 auf dem Balkan und in Rußland

Am 28.3. Kfz.-Marsch
von Berlin über
Dresden
Prag
Wien
Preßburg
Budapest
Arad
Hermannstadt
Kronstadt
in den Raum
Budapest
zum Schutze der
Ölfelder von Ploesti

Nach dem Balkanfeldzug
Bereitstellung zum Ein-
satz "Barbarossa". Ab-
marsch von Ploesti über
Kronstadt
Hermannstadt
Klausenburg
Großwardein
Debrecen
Kaschau
Krakau
Sandomir
in den Versammlungsraum
der Panzergruppe 1
Zamosch, an am 16.6.

Rückverlegung Ende
November mit der
Eisenbahn von Orel
über Brjansk
Smolensk, dann teils
über Witebsk
Dünaburg
Wilna
Kaunas
Insterburg
Königsberg
nach Berlin
oder über Minsk
Baranowitschi
Warschau
nach München

Bis Tula
noch 35 km

Brjansk
Karatschew
Orel
Rylsk
Romny
Priluki
Kiew
Lubny 18.9.
Shitomir
Berditschew 17.7.
Kanew 13.8.
Tscher-kassy 20.8.
Krementschug 13.9.
Swino-gorodka
Alexan-drija
Dnepro-petrowsk 25.8.
Winniza
8.8.
14.8.
Saporoshe 19.8.
Gaissin
Uman 23.7.
10./17.8. Kirowo-grad
Kriwoi Rog 18.8.
19.8. Nikopol

Zamosch 16.6.
Rad-ziechow
Rowno 28.6.
Ostrog 3.7.
Sokal 21.6.
23.6.
Dubno 27.6.
5.7.
Schepe-towka

Weg des geschlossenen RGG
Abweichende Einsätze einzel-
ner Flak-Abteilungen und des
Schützen-Bataillons/RGG
Staatsgrenze zur Sowjetunion
Stalinlinie
Eisenbahntransport

0   50   100   150   200   250   300   km

The RGG is, from June 22, 1941, 12 o'clock noon, incorporated into the Marschgruppe (marching formation) of the 11. Panzer Division (XXXXVIII. Armeekorps) and crosses with it the Bug at Sokal. The Bug river is, since 1939, the border between the Generalgouvernement (military governed) occupied Polish territories and the Soviet Union. At the beginning of the advance the Soviet bunkers on the Sokal heights were effectively shelled by the 'Achtacht' (8.8. cm. gun).

Leutnant Itzen, scouting officer of the 3./RGG, at the still open grave of his Batterie-Chef, Hauptmann Schroder, who had just fallen during the tank battle at Radziechow. What thoughts may move the young officer? Was he thinking of his own death, only a few weeks later? He fell on July 13 near Berditshew. The Ritterkreuz (Knight's Cross) was awarded him posthumously on November 23, 1941.

The three eight-wheelers of the Panzerspah (armored scout cars) Zug proved particularly valuable in the scouting of enemy positions and securing of the flank. They carry, in memory of the Norway campaign, a viking ship as marking.

Briefing on the situation before another scouting venture for the Panzerspäh-Zug.

As best possible we guarded against the dust. Especially helpful were the safety goggles.

Another violent tank battle occurred on June 29, 1941 near Dubno, where the enemy desperately tried to break out, and after the RGG, three days before in the Radziechow-Beresteczko-Lescniow area, had destroyed thirty heavy tanks and eighteen aircraft. The Wehrmacht report of July 7, 1941 states: "In the tank battle near Dubno a Batterie of the Flakregiment "General Göring" under the leadership of Hauptmann Schultz and Leutnant Wilmskotter distinguished itself particularly". (Both belonged to the 2./RGG). In the photo a Panzer KW II, 52 tons. "KW" stands for 'Klement Woroschilow', the Soviet defense commissar.

After Dubno has been cleared by our Panzers and Flak, the Kradschutzen-Kompanie can continue its advance.

Heading towards Ostrong. Again, the batteries of the RGG are in the middle of the advancing leading Panzers.

Past a just knocked-out Soviet tank, between Dubno and Ostrog, a "Achtacht" of the 3./RGG continues its advance.

In Volhynia, the formerly Polish territory, annexed only nine months previously by the Soviet Union, the German troops were greeted as liberators and cordially welcomed. The predominately Roman-Catholic population hoped to have returned to them the churches expropriated by the Soviets and used for grain storage, power generators and repair shops.

Camouflaging of the guns occasionally took on grotesque forms.

Berditschew has been reached. Flak of the RGG guards the bridge crossing.

Heat, dust, and thirst cost strength. Always, only forward, no lengthy rest periods, only the occasional unexpected halt because of a disruption somewhere. The exhaustion grew. Soon, when nature demanded its rights, sleep was possible at any time and in any position.

The vehicle lot of the Regimentsstab after a bombing raid.

When the advance stopped the gunners immediately sought cover behind their tractor.

When there were quarters to be moved into, seldom enough, the chase for "partisans" was on in the evening. The motto was "It is no less shame to get lice, but it is to keep them".

While we advanced further and further east, columns of prisoners, often without guards, moved past us in the opposite direction. In the period from July 27 to August 5, 1941 the Panzergruppe Kleist took more than 27,000 prisoners, and the 17. Armee another 14,000.

Again, a few comrades had to be left behind in foreign soil.

If the point got stuck because fuel and ammunition had run out and supplies could not keep up, a JU 52 dropped emergency supplies of ammunition and a few barrels of fuel.

Called to the front of the assembled Kompanie, some soldiers are given their well deserved awards. On the right, two sergeants of the Panzerspah-Zug whose soldiers still wore the black Panzer beret, which was later no longer used.

In December 1939 the Soviet Union ordered a new tank to be built, the T 34. In June 1941 the first T 34's left the assembly halls and in August the T 34 became our enemy in our sector. It was superior to our Panzers of the time, but was no match for the "Achtacht". Quite a few, as this one, were knocked out by us.

On August 19, 1941 the Dneiper river Soposhe is reached. The Regimentskommandeur with two of his officers at the Dneiper hydropower station "Hydrostanzia", then the third largest dam in the world. Its middle section had been destroyed by the retreating Soviets.

The fall evenings, in particular after it had rained, are noticeably cool. A warm camp fire makes one feel comfortable.

The Regimentsstab is quartered in a Russian farm house. Oberst Conrad and Major Jakobi.

Continuous rains in the autumn caused a period of muddy conditions which lasted until November. The roads lost their foundations. Even this eight-wheeler had problems getting unstuck.

In the bridgehead of Dnjepropetrowsk the Schutzen-Batallion RGG in the sector of the 9. Schutzen-Brigade is put under the command of the Kradschutzen-Batallion 59, since both had the same directives. Its Kommandeur, Major Schmalz, thus has, for the first time, men with the white collar patches under his command. He cannot know that, within a year, he will be wearing these patches himself and later become Divisionskommandeur, even the Kommandierende General (commanding general) of this special unit of the Luftwaffe. On the left in the photo the vehicle of Major Schmalz.

Time and again great numbers of prisoners are brought in. Their fate is now made worse by the beginning of the cold season.

One night a sudden heavy frost hit and all vehicles were frozen tight. It cost a lot of effort to bring them back into driving condition.

Two gunners of the 2 cm.-Batterie have shot down this Rata ('rat') and are inspecting the wreck. The RGG, until October 1941, had already shot down 161 aircraft and knocked-out 324 tanks, destroyed 167 guns and 530 machine-gun positions, blown up 45 bunkers and brought in 11,000 prisoners.

In November a severe cold moved in. Heavy frost brought any movement to a halt. The vehicles could not be started by themselves and without assistance.

The RGG was in action for almost half a year and had reached Orel after incessant fighting, parts of it had even been able to advance to within 35 km. of Tula. Now it was worn out and in need of a rest. It was pulled out and moved by rail from Orel back home. The guns stayed behind.

The Flak-Kampfabzeichen (Flak fighting badge), established in 1941, was awarded to soldiers who had shot down at least five aircraft or (until the establishment of the Luftwaffen Erdkampfabzeichen (Luftwaffe ground fighting badge) and the Seekampfabzeichen (naval fighting badge)) had taken part in three different combat actions against ground and air targets.

The Erdkampfabzeichen of the Luftwaffe, established in 1942 was awarded to soldiers who had proved themselves during fighting at the front in at least three battles on three different days.

The Establishment of the Brigade HG

In the spring of 1942, the RHG was expanded to a brigade with the objective of securing cooperation with paratroop units after landing

For this, the two Flak-Abteilungen in the Paris area and the Schutzen-Batallion, based in Berlin, were moved to Brittany in France,there, the remainder of the second Schutzen-Batallion were already stationed, after returning from Russia for re-deployment.

In the area of Pontivy-Laudeac the units were put together and trained. On July 21, 1942, the establishment of the Brigade HG was completed. It was organized: the Brigadestab (brigade staff) HG, the Flak-Regiment HG with two Flak-Abteilungen, an Artillerie-Abteilungen and a new IV. Abteilungen which took in the Flak-Abteilungen (escort) Batteries, the Schutzen-Regiment HG, containing initially two Schutzen-Bataillon, an Infantrie-(gun) Kompanie and a Panzerjager (tank Hunter) Kompanie and one Panzer Kompanie.

As well, the Wach-Batallion HG and the Musikkorps in Berlin and the reserve detachment (Ersatz-Abteilung HG) continued to exist. However, the until then III. Scheinwerfer (search light) Abteilung/RHG and the reserve Scheinwerfer-Abteilung/RHG, both still deployed for Berlin's air defense, were removed from the RHG and did not become part of the Brigade HG.

The Brigade HG, however, did not see action in battle. In addition to training, security objectives in occupied France were made part of the responsibilities of the Brigade HG.

Participation in fighting actions which were to be counted towards the Erdkampfabzeichen were documented in the personal file, and copied to the new units in case of transfers and temporary postings.

IO. (Sturmgeschütz-) Kp./Fsch.Pz.Rgt. H.G.

Bescheinigung

Dem Uffz. Kanert werden folgende Erdkampftage bescheinigt:

1.	15.7.43	Vernichtung feindlicher Fallschirmjäger bei Lentini
2.	17.7.43	Abwehr feindlicher Angriffe gegen den Simeto im Nahkampf
3.	29.2.44	Vorstoß südlich Cisterna
4.	28.7.44	Erfolgreiche Abwehr eines Panzervorstoßes bei Siennica
5.	29.7.44	Panzerkampf bei Pogorzel
6.	30.7.44	Abwehr eines Panzerangriffes bei Mieszylesie
7.	1.8.44	Angriff auf Struga
8.	19.8.44	Angriff auf Helenowek
9.	13.10.44	Angriff auf Pleine
10.	22.10.44	Abwehr bei Bissnen und Rodebach

(Dienststempel)   gez. Wallhäußer
                    Oberleutnant u. Kompanieführer

On March 1, 1942 the RGG was renamed "Reinforced Regiment (mot) HG". At the same time new sleeve bands with the inscription "Hermann Göring" were introduced.

Ownership certificate: the Gefreiter Gerhard Metzler of the 8./Fsch.Pz. Gren.Rgt. 1 HG is awarded the Erdkampfabzeichen of the Luftwaffe. Signed-Generallt. and Div. Kdr.

At the Brigade staff in Pontivy-/Bretagne. From the left: Leutnant Muller, Hauptmann Vogel, ?, Generalmajor Conrath, Major Bobrowski, Hauptmann Beinhofer, Regierungsober-inspektor (government official) Grosse, Oberleutnant Tilcher, Stabsarzt (staff physician) Dr. von Ondarza.

BESITZZEUGNIS

DEM

Gefreiten Gerhard Metzler
DIENSTGRAD · VOR- UND FAMILIENNAME
8./Fsch.Pz.Gren.Rgt. 1 H. G.
TRUPPENTEIL

VERLEIHE ICH DAS

ERDKAMPFABZEICHEN DER LUFTWAFFE

UNTERSCHRIFT

Generallt.u.Div.Kdr.
DIENSTGRAD

1243　　　MAXIMILIAN VERLAG BERLIN SW 68

Generalmajor Conrath, who on September 4, 1941 as the first member of the RGG received the Ritterkreuz, Major i.G. (im Generalstab'general staff) Werner and Hauptmann Freygang in front of field headquarters of the Brigade in Pontivy.

The Heavy Eisenbahn-(Railroad) Flak-Batterie

At the mobilization, a heavy railroad Flak-Batterie was put into service as 14./RGG. Since it was an experimental unit, it was directly subordinate to the supreme commander of the Luftwaffe and removed for the operational actions of the RGG

After taking over the rolling stock from the Reichsbahn (Reich railroad), the guns (four 10.5 cm. Flak and two 2 cm. Flak), the rest of the equipment, and the training of the men, the first action already took place in September 1939 within the frame of the air defense of Berlin.

The provision with railroad cars suitable for a lengthy fighting stint was insufficient but became greatly improved during the course of the winter through exchanging of the cars and their remodeling to meet troop requirements.

From December 1939 to June 1940, the Batterie was deployed for mobile fighting action in the Rhine river plains and often changed bases. It fought near Bad Krotzingen, Rastatt, Riegel, Saarbrucken, Hugstetten, Karlsruhe and Freiburg. From the Rhine river flats and during the Western Campaign the Batterie fired with great success on bunkers of the Maginot Line. Further areas of action thereafter were Paris, Cherbourg, Bordeaux, Berlin, Stettin, Vienna, Leipzig, Buxtehude, Hamburg and Cologne. Here, the Batterie left the RGG in October 1941.

The Batterie, in roughly two years, had traveled 12,377 kilometers on the rails. The successful Batterie-Chef, Obertleutnant Arnold, his officers and most non-commissioned officers and ranks volunteered to remain with the RGG. These men were replaced and the Batterie was incorporated into the Eisenvahn-Flak-Abteilung 321.

Immediately after taking over the four 10.5 cm. guns during the mobilization, the gunners were trained on this new type of gun. Practice at the railroad freight station Berlin-Reinickendorf.

To use live ammunition for the first time the Batterie went to the Flak firing range Stolpemunde. Here the gunners experienced firing of live rounds from their guns.

A passenger car was re-modeled, thanks to the efforts of the Batterie-Chef, Oberleutnant Arnold, and Hauptwachtmeister (sergeant) Mappes, in do-it-yourself style, to provide a lunch room, class room and recreation facilities.

The Batterie-Chef and the inevitable paper war which was, since the Batterie was an independent unit, particularly voluminous.

"Alarm!" The crews race to their guns. The men still live in cars with the inscription "40 men or 8 horses". These "Landser" (common soldier, GI) sleeping cars were soon exchanged for better cars.

To prevent the incursion of enemy aircraft formations into the Reich territory, fire fights often had to be fought at night.

The Railroad Defense Trains

Many heads of state, Hitler as well, used a train consisting of a salon car, dining car, sleeping car and baggage car for long journeys. Even during peace time, but more so after the beginning of the war, other such trains were put together for some ministers of the Reich. Such special trains were rolling headquarters during the war and thus carried additionally a command car with office, briefing, radio, telephone, and adjutant's compartments, as well as baggage, provision, battery and railroad personnel cars. The Oberbefehlshaber (supreme commander) of the Luftwaffe, Göring, even had two trains available to him which carried the code names "Asien" (Asia) and "Robinson". The train "Asien" even contained a bath done in white tile.

To defend against low level air attacks, each special train was assigned two Flak defense cars which traveled as the first and last cars of the train. On each of these defense cars were two 2 cm. Flak guns-later four barreled. The gun crews had their quarters in the defense cars, they had five compartments with movable bunks. The number of crew of a defense car was seventeen men, consisting of one Zugfuhrer (platoon leader), Halbzugfuhrer (assistant platoon leader), medical corps man, two chief gunners and twelve gunners. Each defense car contained a large freight area in its center which was closed in by three removable doors on each side. This area was used to store ammunition, weapons, equipment and baggage.

In addition to the defense cars the special train, depending on its purpose, consisted of eleven to sixteen cars. It was pulled by two locomotives; for longer trips or steep inclines another locomotive was used to push, at the end of the train.

The Flak crews for the two trains "Zug Reichsmarschall I" and "Zug Reichsmarschall II" as well as for the rains "Zug Fuhrer" and "Zug Reichsaussenminister" (foreign minister) were provided by the RGG. They were named "Wachkommando (guard detachment) Sonderzug (special train) Reichsmarschall I" etc., and were initially contained in the 9./RGG, especially created for this purpose in summer 1939.

The Flak defense cars came in different versions depending on which year they were built. (Baujahr, year built, Lange, length, Hohe, Height, Breite, width)

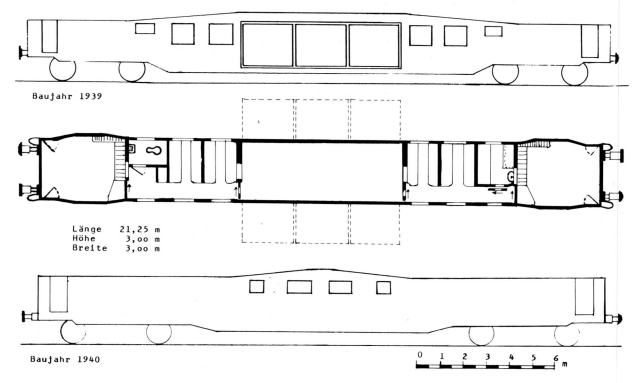

Baujahr 1939

Länge   21,25 m
Höhe   3,00 m
Breite   3,00 m

Baujahr 1940

0 1 2 3 4 5 6
m

Periods of rest were spent in gunnery practice to ensure highest standard of training and readiness.

During an operational journey. Destination is the Spanish border where a meeting between Hitler and Franco is planned. In front Oberleutnant Dieke, commander of the Flak train.

The Flak train "Reichsaussenminister". The gun crew at attention. In action, the crews always wore special clothing, the pilot's flight suit.

The railroad defense trains went to practice with live ammunition at regular intervals. Here a station stop at the Flak shooting range Stolpmunde.

The Fuhrer-Flak-Abteilung (detachment)

At the beginning of the war, the responsibility for guarding the headquarters of the Fuhrer was transferred to the "Fuhrerbegleitkommando" (accompanying squad of the Fuhrer) which was provided by the Infantrie-Regiment Grossdeutschland. It was later renamed "Wach-Kompanie-Fuhrer-Hauptquartier" (guard company) and, still later, "Fuhrer-Begleit-Batallion". To defend against low level air attacks was the task of the 7./RGG, using a 2 cm. self-propelled Flak-Batterie. When mobilized it joined the Fuhrer headquarters under the designation of "7./RGG Flak-Batterie beim Fuhrer-hauptquartier" and was always moved with the Fuhrer headquarters to its different locations.

For the defense of the permanent installation "Wolfschanze" near Rastenburg where the headquarters was often located, the I./Flak-Regiment 604 was deployed. In 1942 it was renamed Fuhrer-Flak-Abteilung and the 7./RGG was incorporated into it. Soon after, the Abteilung joined, as IV. Abteilung, the Flak-Regiment HG and thus received the white collar patches.

When the Fuhrer headquarters was not located at the "Wolfschanze", the troops stationed there for its defense were occasionally used at the eastern front as fighting forces. For instance, the "Kampfgruppe Fuhrer-Flak-Abteilung" fought in the winter battle of Donez from December 1942 to March 1943, centered on the Millerowo area, and sustained heavy losses.

With the establishment of the Fallschirmpanzerkorps HG in October 1944 and the simultaneous formation of the Fuhrer-Flak-Abteilung left the Fallschirm-Panzer-Division HG and was incorporated into the Fuhrer-Flak-Regiment.

The Fuhrer-Begleit-Batallion (including the 7./RGG) on parade at the Berlin-Reinickendorf barracks with the Musik-korps and the musicians of the RGG. There it receives from the hand of the commander of the headquarters, Generalmajor Rommel, its standard.

The Fuhrer-Flak-Abteilung was occasionally deployed away from the ''Wolfschanze'', as at the beginning of 1943 in the area of Millerowo during the winter battle of the Donez. Here the 7./RGG faced hard fighting.

The 7./RGG of the Fuhrer-Flak-Abteilung in action. Chief gunner Unteroffizier Fallowe during daily exercise with his men.

The 2 cm. gun Fallowe was invisible to both visual and photo recognizance of Soviet aircraft.

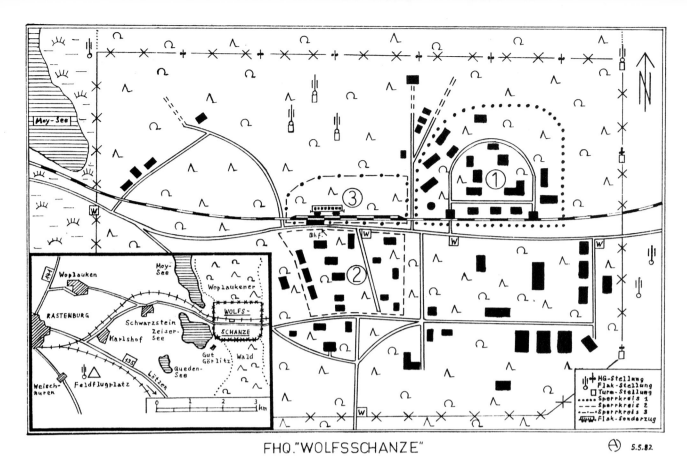

FHQ."WOLFSSCHANZE"   5.5.82

The lay-out of the Fuhrer headquarters near Rastenburg in East Prussia. The three restricted areas were, in addition, surrounded by another fence and heavily secured by guards. Entrance was possible only with special permits. Hitler and the ministers of the Reich, as well as the staffs of the Wehrmacht, had their bunkers in Sperrkreis 1. Sperrkreis 2 contained the lodgings of the guard troops and the railroad installations, the station and the special trains were located in Sperrkreis 3.

'Breakdown of the IV./FlakRgt.HG-FuhrerFlakAbt.' (present deployment)

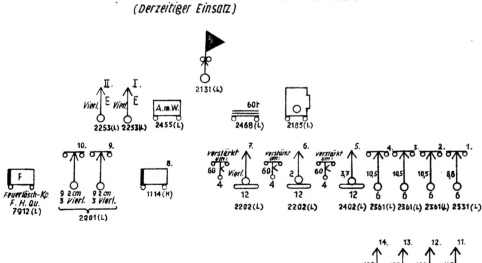

The Establishment of the Division HG

Even during the training phase of the Brigade HG, its deployment plan was expanded that now, beyond the Brigade, a motorized Division, the Division HG, was to be created for the area to the south of Bordeaux. The units of the Brigade HG in the Bretagne were moved to that area. For the establishment and training of the new units, which also took place in Holland and the domestic war theater, as in Berlin, in the camp at Velten, and various training camps, its own staff was created in November 1942. Named the "Gruppe Neuaufstellungen (group from re-establishment) Division HG" (GNDHG), it was based in Mont-de-Marsan with its commander, Oberst Schmalz, provided by the Heer.

The core of the new Division, which became effective on October 17, 1942, were the units of the Brigade HG which still contained many of the men of the previous RGG, who had proven themselves in war and peace. To these were added many specialists from the Heer who had volunteered for the new force and many soldiers of the Luftwaffe ground personnel. The vast majority, however, were volunteers who applied for the new Division from all parts of the Reich.

The Division HG was made up of:

The two Grenadier-Regiments 1 HG 2 HG with three batallions each;
The Jager-Regiments, also with three battalions, previously the Fallschirmjager-Regiment 5;
The Panzer-Regiment HG with two Abteilungen;
The Aufklarungs (reconnaissance) Abteilung HG;
The Flak-Regiment HG with four Abteilungen (including the incorporated Fuhrer-Flak-Abteilung);
The Artillerie-Reigiment HG with four Artillerie-Abteilungen and the V. Sturm-geschutz (assault gun) Abteilung;
The Pionier-Batallion HG;
The Nachrichten (communications) Abteilung HG;
The Sanitats (medical) Abteilung HG;
and, as individual units;
The Divisions-Verpflegungs-Amt HG (office for provisions);
The Backerei (bakery) Kompanie HG;
The Schlachterei (butcher) Kompanie HG;
The Feldgendarmerie (military police) Zug Hg; and
The Feldpostamt HG (field post office).   But, not all of these formations could be immediately provided.

For the Wach-Batallion HG and the Sonderstab (special staff), both in Berlin-Reinickendorf, the Stab.Wach-Regiment HG was formed, and the Ersatz-Batallion HG (replacement battalion) in Holland was expanded to the Ersatz-Regiment HG.

When, on November 11, 1942, German troops moved into the unoccupied parts of France ("Unternehmen (operation) Anton") the units of the division HG in the process of formation or reorganization were also moved there, into the area of Mont-de- Marsan.

('Those who would belong to us must come voluntarily')

After training in the open terrain the Kompanie is on its way home.

It was hard work bringing a heavy infantry gun into position.

The Flak-Regiment also, trained its men using live ammunition even during initial training. In the background Oberstleutnant Hullmann, Kommandeur of the Flak-Regiment HG. To his left the Brigadekommandeur who keeps a sharp eye on the training process.

Young volunteers being trained on the 3.7 cm. Pak (Panzer Abwehr Kanone anti-tank gun).

Ground to ground fire from a self-propelled "Achtacht".

Firing live ammunition from a 10 cm. gun.

Panzer pioneers building a bridge.

Warlike unit exercises closed off training. From the left: Generalmajor Conrath, Hauptmann Rebholz, Oberst Heidmeyer, ?, Hauptmann Schreiber.

Welcome relaxation from the hard training duties was provided by occasional public concerts, as here in the market place of Pontivy under the direction of Obermusikmeister Fries and Oberfeldwebel Bogenschneider.

Inspection of the Panzerpionier-Kompanie. From the left: Hauptmann Tilcher, Major i.G. Werner, Generalmajor Conrath, Oberleutnant Musil.

The Panzer-Regiment HG was primarily formed and trained at the troop training grounds at Munsingen. Here the celebrations of the creation of the Regiment in January 1943. It had, then, Panzer III and Panzer IV tanks.

At the end of 1942, early 1943, the units were gradually moved to the area of Naples. Headquarters of the Division HG became Caserta, north of Naples. Here the Santa Maria barracks.

Well remembered by all; Mt. Vesuvius with its always white, red glowing at night, cloud. Nowadays Vesuvius is extinct.

Highest priority was placed on having the weapons always ready for action, and on controlled care and maintenance under the critical eyes of the Zugfuhrer (platoon leader).

War formation of the Gruppe Neuaufstellungen Division Hermann Göring.

Who did not experience the Neopolitan street vendors?

The Division HG in the African Campaign

In November 1942, the Americans and British landed in Algeria (operation "Torch") to prevent the escape of the retreating German and Italian divisions from Africa. Part of the force, quickly thrown together by the German leadership, which was charged with keeping open Rommel's path of retreat through Tunisia, were also units of the still developing Division HG. As late as November 1942, the Fallschirm-Jager-Regiment 5 (previously Sturm-Regiment Koch), which was incorporated into the Division HG as Jager-Regiment HG, was moved as its first unit, by air to Tunisia. In January 1943 it then followed parts of the Flak-Regiment HG and, weeks later, further units as soon as their formation was completed. They were generally units transported by air in JU 52 and Me 323 Gigant (giant) from Naples or Tripani/Sicily, otherwise by ships, which always carried all heavy weapons and equipment, to Tunis. In Africa they were led by Oberst (as of March 1943 Generalmajor) Schmid, initially as "Kampfgruppe Schmid" later under the designation "Division HG".

The fighting under unaccustomed climatic conditions and against an enemy who was equipped with, then, unimaginable quantities of materiel, in particular tanks, was hard and high in casualties. The HG units fought, in the beginning still attached to various other divisions and later deployed as the complete Division, in the areas of Kairouan, Goubellat, Medjez el Bab, Pont du Fahs, Zaghouan, and Tunis. The Wehrmacht report of April 26, 1943, pointed out that the Division HG with its units in action at the center of the Tunisian front, "distinguished itself through its exemplary fighting spirit and intrepid valor and thwarted the hopes of the enemy for a breakthrough". However, all the bravery did not help. Finally, on May 12, 1943, the troops of the Heeresgruppe Afrika had to capitulate to the overwhelming strength and lay down their arms. Only a few soldiers from the Division HG succeeded in fleeing, in small boats, on an adventurous and extremely dangerous journey to Sicily. Among the 130,000 German soldiers the Allies had "collected in the tied-shut bag" were almost all survivors of the Division HG at the Tunisian front. A particularly great disadvantage for the later recreation of the Division was the fact that the old core, especially the outstandingly trained and battle hardened non commissioned officers, were among the soldiers captured or killed, and were thus greatly missed.

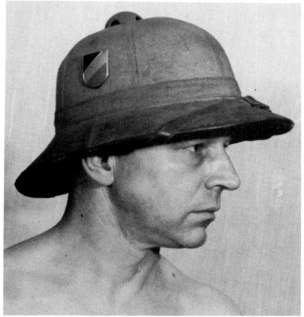

The tropical helmet was an attractive piece of equipment, and who would not have dearly liked to send a photo wearing it to his parents, wife, fiancee, or children? It made one look so photogenic! In practice, however, it was highly inappropriate. It did not protect from bullet injuries and, with the lack of space on the vehicles, many were damaged and became useless. Beginning in the middle of 1943, they were no longer issued.

Let's go! Rations for the march have already been issued. Completely new to us is the canned bread, standard in the navy, which was issued to us.

The order to move has been received. Departure mood in the tent camp which is guarded, in the background, by the Flak.

Near the airport a temporary camp. Renewed waiting for marching orders.

Ready for take-off. The Chef of the 3. Batterie of the Flak Regiment HG in Palermo.

Formations of three aircraft each go non-stop from Palermo to Tunis. Here over Palermo.

Immediately after landing and refueling, the JU's started the return flight to Sicily to bring over more troops. Many of the aircraft, however, did not reach their destination.

Arrival in Tunis. What is next? Waiting, and more waiting.

The gate in Tunis which separates the European and Arab quarters.

Out scouting. Caution-mines!

All movements of the enemy opposite are tensly watched through the "Schere" ('scissors') binoculars.

On the bald, cover-less, Kamelberg (camel hill) which would later be bitterly fought for and was to become our hill of destiny.

(Organization on April 16, 1943)

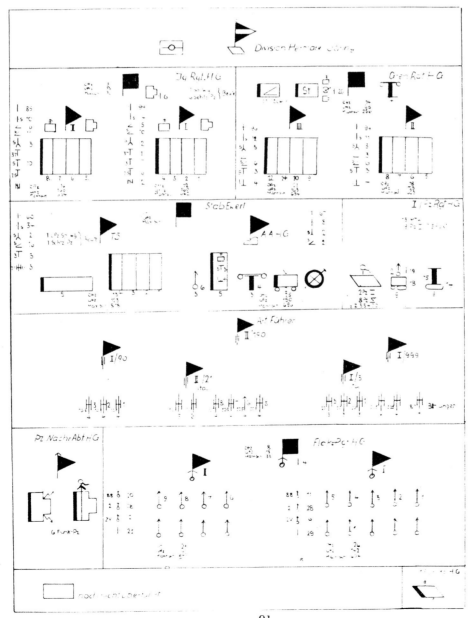

Generalmajor Schmid (Josef) who led the parts of the Division HG in Tunisia and for this received the Ritterkreuz.

Oberfeldwebel Johannes Scheid received the Ritterkreuz as Zugfuhrer in the Grenadier-Regiment HG for his brave actions in Tunisia. It was presented to him in a ceremonial form in French captivity.

The physical well-being also had to be looked after. In the shde of an olive tree the field kitchen has started its activity.

The Leutnant issues last orders to his squad leaders.

Again, a few comrades had to be left behind. As a last salute only a wreath of palm leaves decorated the graves.

Break down of the forces of the Heeresgruppe (army group) Tunis in April 1943.

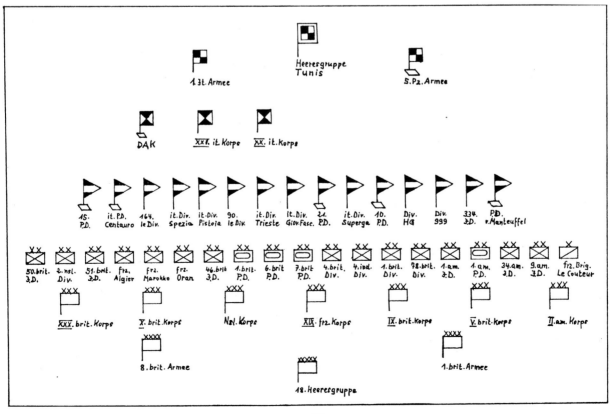

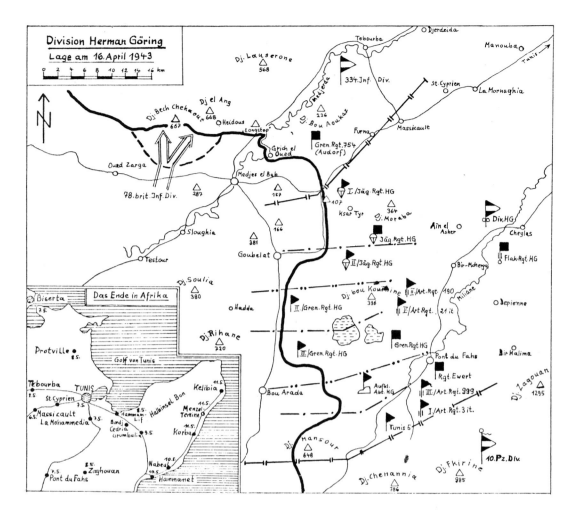

Map of the Situation on April 16, 1943.

'Interim Award Certificate. In the name of the Führer and Commander-in-chief, I award to the Gefr. Freihalter, Division Hermann Göring, the Iron Cross 2nd Class. Battle headquarters, May 11, 1943. signed: Generaleutnant'

Because of the catastrophic supply situation, the troops were missing many items, including the award forms for decorations. Finally, typed interim award certificates in simple format had to be issued.

The German soldier in Africa fought bravely to the last bullet and did not lose his determination, not even when the enemy dropped many leaflets on the German positions. One of these read: "Poor little Afrikacorps, can't go back, can't go ahead anymore. Before you is hell, behind you the sea, you'll never see your homeland again".

On May 12, 1943, the fighting was over. Among the countless prisoners were almost all survivors of the Division HG. In the photo one of the long columns marching to the central receiving camp.

The dead of the 'bridgehead Tunisia' were initially buried in six temporary cemeteries and only later moved to the German military cemetery in Bordj Cedria. In the earlier cemetery in La Mornaghia, 322 fallen from the Division HG alone were interred, most of them from the Jager-Regiment HG.

The 8,625 German soldiers who fell in Tunisia, among them some 400 from the Division HG, have found their final rest since 1977 in the German military cemetery at Bordj Cedria, located on Route 1, 25 kilometers east of Tunis. They rest in bone chambers made of stone, whose surfaces have engraved on them the names of the dead behind their walls.

Most of the prisoners of the Division HG were brought to America. In the officers' camp of Concordia in Kansas they met again, among others, and are seen in the photo: Oberstleutnant Funck, Major i.G. Werner, Major Neubauer, Captains Dieke, Huhnfeld and Tilcher, Regierungsober Inspector Grosse, and the Lieutenants Krause and Mohn.

The Re-establishment of the Division HG

The gaps caused by the loss of the units in Tunisia were closed by immediate new replacements, and every effort was made in all haste to combine these with the units in southern France, which had not made it to Africa, into a new division, quickly ready for action.

The staff of the Gruppe Neuaufstellungen Division HG was moved at the beginning of May 1943 from Mont-de-Marsan temporarily to the troop excercise grounds Munsingen, and two weeks later to Santa Maria/Capua Vetere, located north of Naples. There it formed on May 21, 1943 the staff of the Division (mot.trop.'motorized tropical) HG and the staff of the Grenadier-Brigade z.b.V. HG.

At the same time all units formed for the new Division were relocated from southern France, Holland, Berlin and Velten, as well as from various exercise grounds in the Reich, to Italy. They moved into bivouac camps in the area of Naples-Caserta-Capua and trained in fighting as units. This was in preparation of the defense against an enemy landing operation which was becoming likely.

Following the evaluation of its intelligence information, the German leadership had to expect a landing of the Allies in Sicily. The Division (mot. trop.) HG was then moved, piece by piece, until the end of June 1943 to the island of Sicily and made ready for action in the area of Caltagirone.

The final form of the Division HG was meant to be that of a Panzer-Division of the Wehrmacht, and the only one of the Luftwaffe, was already involved in heavy fighting against the invasion forces in Sicily.

The organization of the Panzer-Division HG changed (as was the case with other units) later repeatedly. The Division included on January 1, 1944 the following branches:

Stab (staff) Panzer-Division HG-Division commander: Generalleutnant Conrath; Brigadestab z.b.V. HG-Brigade commander: Oberst Schmalz; Panzer-Regiment HG with two Panzer-and one assault gun detachment; Panzergrenadier-Regiment 1 HG with two Panzer-and one assault gun detachment; Panzergrenadier-Regiment I HG with two batallions, Panzergrenadier-Regiment 2 with two batallions; Panzeraufklarungs (reconnaissance) Abteilung HG; Flak-Regiment HG with three detachments and the attached Fuhrer-Flak-Abteilung; Panzerartillerie-Regiment HG with four detachments; Panzerpionier-Bataillon HG; Panzernachrichten (communications) Abteilung HG; Feldersatz (replacement) Bataillon HG; Divisionskampfschule HG (divisional war school); I. Nachschub (supplies) Abteilung HG; II. Instandsetzungs (maintenance) Abteilung HG; Verwaltungsttruppen (administration) HG; Sanitats (medical) Abteilung HG; Feldpostamt (field post office) HG. In addition, the following units and offices existed: Wach (guard) Regiment HG (in April 1944 remodeled to the Begleit (escort) Regiment HG) and liaison staff both in Berlin-Reinickendorf; Ersatz-und Ausbildungs (replacement and training) Regiment HG in Holland; Leichtkranken (non-critically ill) hospital and recuperation station in San Martino di Castrozza/Dolomites.

The Divisionkommandeur inspects, in March 1943, the re-established Flak-Regiment at the Flak range Deep.

The Battle for Sicily

On July 10, 1943, the American 7th Army and the British 8th Army landed in Sicily (operation "Husky"). The Panzer-Division HG and the 15. Panzer-grenadier-Division, quickly established from the 'Africa stop' (stalled battalions on their way to Africa, soldiers returning from home leave or recuperating) were initially the only major German units available to assist the hastily deployed XIV Panzerkorps (General Hube) in the defense of the island. The Italian troops, the 6th Army (General Guzzoni) with ten divisions, retreated, with few exceptions, without fighting to the mainland or, those whose home was Sicily, simply dissolved. Portions of the 1. Fallschirm-Jager-Division arrived on July 12 from southern France as reinforcements, parachuting into the area of Cataniz and the Simeto bridge, and on July 15 the 29. Panzergrenadier-Division arrived from Calabria.

The Panzer-Division HG fought its way, divided into two battle groups led by Generlleutnant Conrath and Oberst Schmalz respectively, from Gela, Caltagrione, Catania, Acireale, Adrano (past the Aetna range) and via Randazzo-Taormina back to Messina. At the end of the battle, it was shipped by ferries, landing craft and attack craft across the straits of Messina to Calabria.

After overcoming shock in a few units during the first two days of fighting, the troops caught themselves quickly and showed, despite many adversities, outstanding results in the fighting. The Division took a decisive part in the delaying defense of Sicily which allowed the total return of even the heavy war materiel. During the 38-day battle for the island, in the greatest heat and with an, until then unimaginable, superiority of materiel and air power of the Allies, the troops deployed here showed respectable abilities. The report of the Wehrmacht of August 18, 1943 states with appreciation "Leadership and troops have achieved results which will be entered into the annals of war equally as would a victorious attack". The enemy as well, General Eisenhower, judged in his book 'Crusade in Europe':

"Along the great, saw tooth like range centered on the Aetna, the German troops still fought adroitly and with fierce determination. The Panzer and paratroop units deployed here belonged to the best we encountered throughout the entire war, and each position could only be taken when its defense was totally destroyed".

The Panzer-Division HG then pulled back through Calabria and Apulia in the hope of finding time and occasion for the greatly needed refurbishing in the area of Naples. It would, however, turn out differently.

This is the small fishing village of Canitello near Villa San Giovanni, where the docking facilities for the ferries of the Division HG were installed. On the other side of the approximately six kilometer wide straits of Messina lies Sicily.

98

A Panzer IV of the 7./Panzer-Regiment HG is loaded to be ferried across.

Those who did not arrive in Sicily by troop transport but traveled individually or after leave from Berlin to the South, followed the route indicated on this railroad sign.

The port of Messina was heavily damaged by British and American air attacks and became almost unusable because of the sunken ships.

The troops had set up under gnarled olive trees and between cactus growths and were full of anticipation.

Under a mighty olive tree the divisional supply chief had opened his office. Together with his assistants he ensured that the soldier received all necessities for life in the field. From the left: Stabsfeldwebel Vaupel, Oberfeldzahlmeister Klopfel, Oberzahlmeister haussler, Stabszahlmeister Otte (last three ranks refer to functions of supply and pay).

In the supply camps, food was prepared, vehicles, weapons and equipment repaired and provisions stored.

The narrow, steep, and curved roads and tight village streets required high abilities from the Panzer drivers. Here a Panzer of the 1./Panzer-Regiment HG during a stop on a mountain road.

Songs of the Sirens

'To a Grenadier of the Hermann Göring Division'

Welcome. You are back in action. For the third time here at the southern front. First Sicily, then from Salerno to Volturno and now in the rocks and shot-up mountain positions of central Italy. We already know each other well. The old Hermann Göringers we already caught in Tunisia. They are now sitting in our homeland enjoying life...You are back in action and probably wondering why there is now so much talk of being a prisoner of war. Suddenly they tell you that we Anglo-Americans mistreat prisoners of war. The reason for this is obvious: In case you find yourself in a hopeless situation they want to prevent you from surrendering. Instead they offer you the hero's death. No, you may not let yourself be taken prisoner although the war is lost for Germany, although the hero's death of German soldiers can only have one purpose: To prolong the war for a few months.Or, maybe the 'secret weapons' will change the situation? The Do-mechanism, the compressed air bomb, the rocket launcher, the 6-engine bomber, and whatever else, are no longer secret weapons.

You are a soldier. Ask yourself honestly whether it still makes any sense to sacrifice yourself. You have the strength to accept reality, the duty to recognize the hard truth, and the right to draw the proper conclusion. German soldier: We promise you neither utopia nor heaven on earth if you are taken prisoner of war. But you can definitely count on the following facts:

1. Fair treatment, as a courageous soldier deserves. The rank of a prisoner will be recognized. Your own comrades will be your immediate superiors.
2. Good provisions. Many of your comrades are surprised how well they are being fed by us. We rightfully have the best fed army in the world. (Many a Landser prefers the German Kommisbrot to our white bread, but no-one has yet complained about our coffee and the preparation of our meals).
3. First Class Hospital Care for the wounded and the ill. According to the Geneva Convention prisoners receive the same hospital care as our own troops.
4. Opportunity to Write. You can write three letters and four post cards to your home per month. Postal connections are fast and reliable. You can receive letters and also parcels.
5. Pay. According to the Geneva convention the prisoner retains the right to his pay. For possible voluntary work you will, of course, be paid. With the money you receive you can buy various goods and luxury items.
6. Further Education. Should the war continue for some time yet, you will probably find occasion to participate in various educational and academic courses which will be organized by the prisoners of war themselves.
And, of course, you will return home at the end of the war.

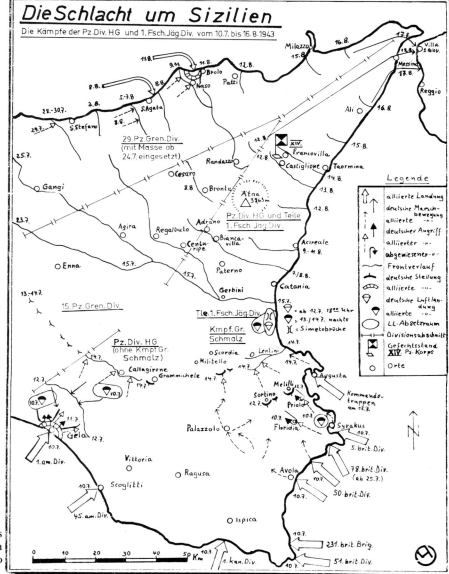

Die Schlacht um Sizilien
Die Kämpfe der Pz.Div. HG und 1.Fsch.Jäg.Div. vom 10.7. bis 16.8.1943

The Battle for Sicily.

Legend:
Allied Landing
German Movement
Allied Movement
German Attack
Allied Attack
Repelled Attack
Location of Front
German Position
Allied Position
German Parajump
Allied Parajump
Airborne Landing Area
Divisional Boundary
Battle Command XIV Pz.-
Korps
Villages

Legende

	alliierte Landung
	deutsche Marsch-bewegung
	alliierte " "
	deutscher Angriff
	alliierter " "
	abgewiesener " "
	Frontverlauf
	deutsche Stellung
	alliierte " "
	deutsche Luftlan-dung
	alliierte " "
	LL-Absetzraum
	Divisionsabschnitt
XIV	Gefechtsstand XIV Pz.-Korps
o	Orte

The 2 cm.-Flak has proven itself as an exceptional weapon also during ground fighting.

The divisional commander inspects the defensive positions, here a Pak position.

The enemy propaganda effort was extremely active and tried, above all, to weaken our fighting spirit through leaflets, but without success.

DEUTSCHE SOLDATEN IN SIZILIEN

EUERE LAGE IST HOFFNUNGSLOS —

UND IHR WISST ES ! !

Am 10. Juli landeten wir mit 3.000 Schiffen, heute sind zwei Drittel der Insel in unserem Besitz.

Ihr steht einer erdrueckenden Uebermacht gegenueber !

Wo Ihr eine Division habt, da haben wir eine **Armee.**

SCHAUT AUF ZUM HIMMEL ! Der Luftraum gehoert uns.

SCHAUT AUF DAS MEER ! Nur ein halbes Prozent der gesamten Schiffstonnage, die wir bei der Landung einsetzten, ging verloren. Der Ozean und das Mittelmeer gehoeren uns. Unsere Verbindungen sind ueberall offen. Euere Verbindungen sind hoechst gefaehrdet und unter ständigem Bombenhagel.

SCHAUT AUF EUERE GEGNER ! Ihr steht denselben Armeen gegenueber, die das Afrika-Korps, die alte Hermann Goering Division und die 15. Panzerdivision in Tunesien vernichtet haben. Euere Gegner haben die Kampferfahrung des ganzen afrikanischen Feldzuges. Sie haben hinter sich die gewaltigen Reserven der Vereinigten Nationen an Menschen und Material.

SCHAUT AUF EUERE VERVUENDTEN ! Der italienische Soldat will nicht mehr fuer Mussolini und den Faschismus kaempfen, auch nicht auf italienischem Boden. Das ganze italienische Volk will den Frieden.

Sogar unter Eueren eigenen Truppen sind Gegner. Ihr wisst selbst, wieviele Soldaten aus den unterworfenen Ge-

bieten in Eueren Reihen stehen. Das sind keine Deutschen. Sie wollen nicht fuer Deutschland kaempfen.

SCHAUT AUF EUER EIGENES LOS ! Auf verlorenem Posten werdet Ihr sinnlos geopfert. Wieder wird man Euch sagen, Ihr muesst bis zum letzten Mann kaempfen, um Zeit zu gewinnen. Aber wie tapfer Ihr auch kaempft; Italien ist verloren ! Wollt Ihr fuer ein Italien sterben, das selbst der italienische Soldat nicht mehr verteidigen will.

SCHAUT 10 WOCHEN ZURUECK ! Euere Kameraden in Tunesien waren in derselben Lage. Sie haben tapfer gekaempft. Das Schicksal hat gegen sie entschieden. 248.000 Gefangene wurden gemacht, mehr als die Haelfte davon waren Deutsche. Sie haben die ehrenvolle Kapitulation dem sinnlosen Untergang vorgezogen. Als Kriegsgefangene unter dem Schutz der Genfer Konvention, wissen sie heute, dass wir Ihre Haltung achten. Sie werden die Heimat wiedersehen !

SCHAUT IN DIE ZUKUNFT ! Zwei Wege stehen Euch offen. Der eine fuehrt in die sichere Vernichtung und in den Tod, der andere fuehrt ueber unsere Linien.

Es ist der einzige Weg zurueck ins Leben, zurueck in Euere Heimat.

Alle Einheiten der Armeen der Vereinigten Nationen in Sizilien haben den ausdruecklichen Befehl, alle deutschen Soldaten, die dieses Flugblatt vorzeigen, oder sonst ein klares Zeichen geben und die ohne Waffen zu uns herueber kommen, in Sicherheit zu bringen.

SAFE CONDUCT	Deutsche Uebersetzung PASSIERSCHEIN
I surrender and place myself under the protection of the Geneva Convention for prisoners of war.	Ich ergebe mich und stelle mich unter den Schutz der Genfer Konvention fuer Kriegsgefangene.

German soldiers in Sicily: Your situation is hopeless-and you know it!! On July 10 we landed with 3,000 boats, today two thirds of the island is in our hands. You are facing an overwhelming superior force! Where you have a division we have an army. Look up to the sky! The air space belongs to us. Look at the sea! Only a half percent of the total ship tonnage we deployed for the landing was lost. The ocean and the Mediterranean belong to us. Our supply lines are open everywhere. Your supply lines are greatly endangered and under a constant hail of bombs. Look at your enemy! You are facing the same armies that annihilated the Afrika-Korps, the old Hermann Göring Division and the 15. Panzer-Duvision in Tunisia. Your enemies have the battle experience of the whole African campaign. They have behind them the vast stores of the United Nations of people and materiel. Look at your allies! The Italian soldier does not want to fight anymore for Mussolini and fascism, not even on Italian soil. The whole Italian people want peace. Even among your own troops are enemies. You know yourselves how many soldiers from the occupied territories are in your ranks. They are not Germans. They do not want to fight for Germany. Look at your own fate! In a lost situation you are being senselessly sacrificed. Again you will be told that you must fight to the last man to gain time. But, however courageously you may fight, Italy is lost! Do you want to die for an Italy which even the Italian soldier will no longer defend? Look back ten weeks! Your comrades in Tunisia were in the same situation. They fought valiantly. Fate decided against them. 248,000 were taken prisoner, more than half of them were Germans. They preferred honorable capitulation to senseless death. As prisoners of war under the protection of the Geneva Conventions they know today that we respect their attitude. They will see their home again! Look into the future! Two paths are open to you. One leads to certain annihilation and to death, the other leads to our lines. It is the only way back into life, back to your homes. All units of the armies of the United Nations in Sicily have the express order to take to safety all German soldiers who show this leaflet, or who give another clear sign, and who cross over to us without weapons.

The Panzers, guns, vehicles and other equipment found well camouflaged spots in the olive stands.

Leutnant Spitzbarth leads an Infantrie-Kompanie put together from crews whose Panzers have been knocked out. When he had to scout cross-country, the mule was more efficient than a vehicle.

Sicily is left behind, the mainland comes closer.

Protected by a four-barrel Flak, the return to the mainland after 38 days of fighting is accomplished by ferry.

The fallen on the island of Sicily. 4,561 fallen German soldiers rest in the German soldier cemetery of Motta San Anastasia near Misterbianco, located 9 km. west of Catania.

The Battle of Salerno and the Fighting in Southern Italy

A replenishment of the Panzer-Division HG did not come about, however. Without having had any rest the Division, together with the 15. Panzergrenadier-Division and the 16. Panzer-Division, had to make ready again to defend, on September 9, 1943, against the American 5th Army which had landed during the previous night in the bay of Salerno (operation 'Avalanche'). The heat, air superiority and devastating fire from the ships' artillery demanded the last from the troops during the seven-day battle of Salerno. Added to this was the defection of the previous ally with all the resultant symptoms and uncertainties ('case axis'). But at least now clear relations with the Italians had been created, their uncertain attitude had been watched for weeks already with great concern.

Because of the disproportionate relative strength the landing could not be prevented. The fighting against the advancing Americans continued during a slow defensive retreat. The Panzer-Division HG was still attached to the XIV. Panzerkorps which was now subordinate to the newly established 10. Armee (General von Vietinghoff). The locations of battle of the Panzer-Division HG are here connected to the names Volturno River, Mingnano and Garigliano River.

Only at the beginning of November 1943 could the first parts of the Panzer-Division HG be pulled out of the fighting. After almost four months of uninterrupted combat they moved to resting positions at the foot of the Lepine Mountains in Frosinone province. Some units, however, in particular those of the Flak-Regiment HG and the Panzerartillerie-Regiment HG, attached to other divisions, had to continue, some even until January 1944, the heavy defensive fighting.

The battle of Salerno was marked particularly by the impact of the enemy ships' artillery which mercilessly battered everything.

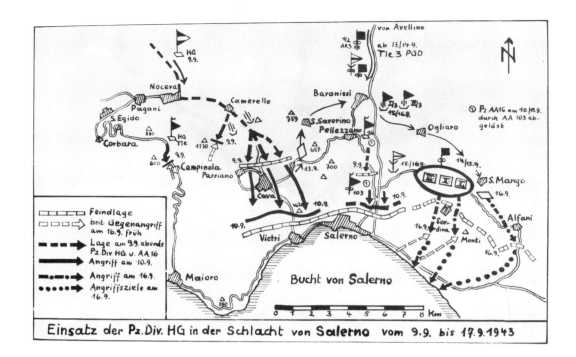

Einsatz der Pz.Div. HG in der Schlacht von **Salerno** vom 9.9. bis 17.9.1943

Action of the Pz.Div. HG during the battle of Salerno from 9/9 to 9/17, 1943.
Enemy Lines
British Counter Offensive early 9/16
Situation on 9/9 in the evening Pz. div HG and AA16
Attack on 9/10
Attack on 9/16
Objects of the attack on 9/16

The divisional commander inspecting the front.

Unteroffizier Kanert with the crew of his assault gun (Sturmgeschutz III) near Castrocielo/Cassino. To the left of Kanert: Haertle; to his right: Jager and Gottschalk.

The grenadiers defended each position doggedly and with tenacity. Fighting in the built up areas was especially tough.

The fallen of the Panzer-Division who were buried in the temporary soldiers' cemetery at Roccasecca later found their last resting place in the German soldiers' cemetery Cairo near Cassino.

Grenadiers of the Panzer-Division HG who were killed during a fighter bomber attack on a truck were given a temporary grave next to the road Via Casilina. They were later moved to the soldiers' cemetery Cairo near Cassino.

The old Roman road Via Appia which leads from Littorna (now Latina) 40 kilometers in a straight line to Terracina was one of the main arteries for supplies. It was, as was the Via Casilina, the favorite target of enemy fighter bombers. Thus the roads were overwhelmingly empty during the day since the fighter bombers attacked every vehicle and fired on any movement, even on motorcyclists.

In Cassino, two roads important to troop movements split off the Via Casilina. Signs of the units of the Panzer-Division HG give directions.

In the inner yard of the German soldiers' cemetery on Sicily lies, stretched out on a pedestal, the figure of a dying youth, a bronze statue of great power. Light and shadow, symbols of life and death, meet here.

Saving of the Cultural Treasures of the Monastery Monte Cassino

In October 1943 a new defensive position, running straight across the Apenine peninsula, the 'Gustav' line, was established before which the advance of the Allies at the narrowest span of the peninsula was to be halted. This defensive line cut the strategically important main road between Naples and Rome, the Roman era Via Casilina, near Cassino. The city of Cassino, at the foot of the monastery Monte Cassino, was to be included in the defensive zone.

When the Kommandeur of the Instandsetzungs-Abteilung HG (repair unit) Oberst-leutnant Schlegel of Vienna, who was interested and knowledgeable in the arts, learned of this, he realized that the patriarchal monastery of Western monasticism, the Benedictine Abbey Monte Cassino, would be in extreme danger of destruction during the expected fighting. Schlegel called on the 80 year old abbot of the monastery, Archabbot-Bishop Gregorius Diamare, and suggested, in view of the threatening destruction of the monastery, to have the many irreplaceable treasures of art, with his help, moved to safety. The abbot initially refused, but after another intervention by Schlegel, realized the danger. Thus Schlegel, who had acted completely on his own and without the knowledge of his superiors, was able to set in motion a cultural act of truly historical significance.

With the help of his soldiers some 80,000 volumes of the valuable library, some 1200 irreplaceable manuscripts, i.e. those of Cicero, Horace, Virgil and Seneca, countless unique scrolls and other archival materials, many valuable paintings and works of art and similar valuables, in particular also the relics of the holy Benedict, founder of the Benedictine order, could be removed from the danger zone.

However, what only became known during the rescue operation was that many world renown paintings of European masters, destined for an arts exhibit in Naples at the time when the peninsula was still spared from the war, were also stored in the monastery.

Schlegel had the property of the monastery moved to Rome into the control of the church. The paintings, owned by the Italian state and once destined for an arts exhibit, were initially brought to the castle of Spoleto, some 150 kilometers north of Rome, later handed over in Rome to the responsible authorities.

120 trucks were needed to bring all these treasures from the endangered monastery area. The charging of the soldiers with duties which were not essential to the war effort, the use of vehicles for non-military purposes were offenses which were subject to severe penalties by courts martial, often the death penalty. Schlegel carried the responsibility initially alone. Later, when his divisional commander , Generalleutnant Conrath, learned of it he covered up Schlegel's arbitrary act and supported the rescue operation himself.

When it was complete, a service to give thanks took place in magnificent basilica of the monastery on November 1, 1943 for all involved in the rescue. At the end of the mass Schlegel received, from the hand of the abbot, a document of thanks written on precious parchment.

On February 15, 1944, the monastery was turned into rubble and ashes by a hail of 550 tons of bombs dropped from more than 250 Allied bombers. But the treasures of art were in safety. Among them were also paintings by Titian, Raphael, Tintoretto, Ghirlandajo, Brueghel and Leonardo da Vinci, pictures which belong to the world.

A few months later Schlegel was critically wounded during a fighter bomber attack and lost a leg. After the war he had to endure many humiliations in his home city of Vienna. Based on wrongful accusations he was sentenced to prison and his family was ejected from their home. After finally achieving his rehabilitation, he later became a member of the Vienna city council. After a long and severe illness, he died on August 8, 1958 in Vienna.

In October 1969, a memorial plaque, donated by his former divisional comrades, was unveiled at the home where he had lived and died, with the participation of the Austrian Federal Armed Forces.

Picture bottom right: The 81 year old abbot-bishop Diamare hands over the valued treasure of the monastery, the relics of the holy Benedict and his sister, the holy Scholastica, to remove them from the war zone.

The patriarchal monastery of the Benedictines, Monte Cassino, atop the 517 meter monastery mountain before its destruction on February 15, 1944 by Allied air attacks. To its left the 'Hill 434', later bitterly fought for, and known to all Cassino fighters as 'gallows mountain' because of the tower for the cable car rope. It was destroyed and has not been rebuilt.

During a Thanksgiving service on November 1, 1943, the abbot presented the savior of the arts treasures a hand painted document which, translated, has the following text: "In the name of our lord Jesus Christ. The noble and beloved military tribune Julius Schlegel, who saved monks and treasures of Monte Cassino with much effort and devotion, is thanked from the bottom of their hearts by the Cassinensers and they plead with the Lord for his welfare. Monte Cassino, the 1st November 1943. Gregorius Diamore O.S.B., Bishop and Abbot of Monte Cassino."

The German soldiers' cemetery Cairo near Monte Cassino. Here rest 20,047 German soldiers, among them those of the Panzer-Division HG, who fell in southern Italy, during the fighting in the retreat there and in the Cassino area, and in particular during the Battle of Salerno.

The last of the works of art
enter the Castle of the Angels.

Oberstleutnant Julius Schlegel.

In front of the Castle of the Angels Oberstleutnant Schlegel
reports to the abbot-primate of the abbey San Anselmo, Fidelis
von Stotzingen, with the hand-over of the last transport, the
successful completion of the rescue operation. next to Schlegel his
adjutant, Leutnant Raab; in the center, officers from the staff of
Field Marshall Kesselring who are deployed as experts for the
securing of endangered works of art.

The Battle for the Beach-head of Anzio-Nettuno

While most of the Panzer-Division HG was established for replenishment between Frosione and Gaeta, the Monte Sammucro, and the Monte Troccio in January 1944, then near Castelforte and at the Rapido river. During this time, our division, effective January 6, 1944, was renamed Fallschirm-Panzer-Division HG. It continued to be deployed, however, as a Panzer-Division in the framework of the army.

On January 22, 1944, at two a.m., the American VI corps landed unexpectedly near Anzio and Nettuno (operation shingle). It was the Allied goal, with simultaneous start of the first battle for Cassino, through fast advance to the Alban Mountains and capture of Rome, to cut the two state roads 6 Via Casilina, and 7, Via Appia, vital supply routes for the German troops fighting at Cassino and to prevent their escape.

The German leadership, with units quickly put together and among them the Fallschirm-Panzer-Division HG, laid a ring around the landing areas. A see-saw battle followed which lasted until May 3, 1944. Leading on the German side was the 14. Armee (Generaloberst von Mackensen) with the I.Fallschirmkorps and the LXXVI. Panzerkorps. The Fallschirm-Panzer-Division HG was mainly deployed on the left wing of the beach-head, approximately from Cisterna to the coast near Borgo Sabotino. The fighting was marked by detrimental weather conditions, in particular unending downpours and by muddy conditions in the former Pontine Marshes, and by enemy materiel superiority. Again, heavy fire from ships' artillery and unchallenged Allied air supremacy set the tone of the battle.

After the failure of another German counteroffensive to eliminate the beach-head during the first days of May, the front initially became calm. At this time the Fallschirm-Panzer-Division HG was removed as completely worn out, and found itself since the beginning of March 1944 in the area of Lucca-Pisa-Livorno, in beautiful Tuscany, for refitting, with an additional charge: securing the coast.

On April 15, 1944 the divisional commander, Generalleutnant Conrath, received orders to another posting. His successor was the, until then, commander of the Fallschirm-Panzergrenadier-Brigade z.b.V., Oberst Schmalz, with simultaneous promotion to major general.

Large units on January 28, 1944. Until the end of February additionally deployed a the beach-head: LXXVI. Panzerkorps, 29. Panzergrenadier-Division and 114. Jager-Division.

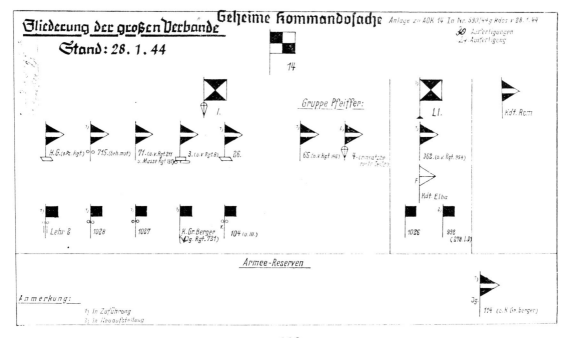

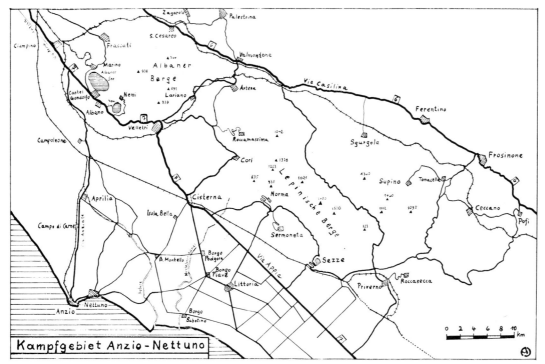

Kampfgebiet Anzio-Nettuno

Despite devastating artillery fire, especially from the sea, despite daily area bombardment and immense deployment of materiel which we had nothing to counter with, the Allies were unable to achieve the planned quick break-through to Rome. In addition there was terrain completely unsuitable for tanks and the adverse weather; rainfalls turned the ground into swamp. The battle came to a standstill at the beginning of March in the mud. It was a battle of attrition with high losses on both sides.

What the enemy was unable to achieve with weapons he tried through psychological influence, mostly through massive drops of leaflets.

Allied Landing near Rome! Strong divisions of the 5th Army with tanks and heavy artillery are now positioned between you and Rome. The main battle line in the south has been by-passed. If you turn north or south, you will have the enemy in front or at your back. The battle in the south will become a battle of circlement. Under the cover of heavy units of the fleet and superior Allied air force an pitiless ring closes. With one blow your situation has become desperate. Any attempt of relief or break-out can only lead to bloody losses as at Stalingrad. The immediate future will bring you a bloody, senseless Stalingrad, or a cruel run of the gauntlet under the hail of the Anglo-American aircraft, or an orderly surrender as at Tunis.

Alliierte Landung bei ROM!

Starke Divisionen der 5. Armee mit Panzern und schwerer Artillerie stehen jetzt zwischen Dir und Rom. Die HKl. im Sueden ist umgangen. Ob Du Dich nach Norden oder Sueden wendest, Du hast den Feind vor Dir und im Ruecken. Die Schlacht im Sueden wird zur Kesselschlacht.

Unter dem Schutz schwerer Flotteneinheiten und der ueberlegenen alliierten Luftwaffe schliesst sich ein unerbittlicher Ring. Mit einem Schlag ist Deine Lage eine verzweifelte geworden. Jeder Versuch der Entsetzung oder des Ausbrechens kann nur zu blutigen Verlusten fuchren, wie bei Stalingrad.

Die unmittelbare Zukunft bringt Dir ein blutiges, sinnloses Stalingrad, oder ein grausames Spiessrutenlaufen unter dem Hagel der anglo-amerikanischen Flieger, oder eine geordnete Uebergabe wie bei Tunis.

The battlefield in the section of the Fallschirm-Panzer-Division HG. These are the drained ditches, was not well suited for tank activity. Farm buildings were scattered throughout the terrain. The treeless plain did not offer any cover to the troops. Fox holes and trenches soon filled with water.

114

The German fighter defense was missing. The available anti-aircraft forces were too small in number to effectively deter the constant air attacks.

Bombs of the largest caliber left gigantic craters as here in Velletri.

When other connections failed, communications were kept open through motorcycle dispatch riders.

Death reaped a rich harvest. The fallen were initially buried in a war cemetery near Rome and later moved to the German war cemetery Pomezia.

After the fighting at the bridge-head Anzio-Nettuno, Major Fitz, commander of the I./Fallschirm-Panzergrenadier-Regiment 1 HG, was awarded, as the 511th soldier of the armed forces, the Oak Leaves to the Knight's Cross. The Tyrolean Fitz was one of the bravest officers of our division, an old trooper, who new how to capture the hearts of his men. He suffered nine wounds whose after-effects led to his death in 1977.

The top leadership: Divisionkommandeur Generalmajor Conrath (center), Brigadekommandeur Oberst Schmalz, and Oberstleutnant i.G. von Bergengruen.

The Fight for Rome and the Retreat to the Arno

The period of rest in Tuscany was used to replenish the units, to repair weapons, equipment and vehicles, clothing and kits, and to grant the men some relaxation after the hardships of the last months. But even after only a few weeks the Fallschirm-Panzer-Division HG had to return to the action prematurely and unexpectedly.

The Allies started, after having greatly reinforced their troops during the previous week, a major offensive from the bridge-head of Anzio-Nettuno with the objective of finally reaching Rome. During the first thrust Cisterna and also Velletri were lost. When this major offensive became apparent the Fallschirm-Panzer-Division HG was again ordered to the front at forced march speed on May 23, 1944. For the sake of speed, marching was ordered also during the day, a fatal mistake by the leadership. On the roughly 350 kilometer long route the units, in particular also the Fallschirm-Panzerartillerie-Regiment HG, already suffered such high losses from low level air attacks that they reached the front with greatly reduced combat readiness. They were still able, partly through counterattacks, to sufficiently delay the breakthrough of the Americans near Valmontone which had as its goal the cutting of the Via Casilina. The losses, however, were frightening.

The armed forces report of June 4, 1944 stated that, "near Valmontone the Fallschirm-Panzer-Division HG fought heroically".

On June 4. 1944, the Americans marched into Rome, more than four months later than had been planned.

The further rear-guard actions north of Rome along the valley of the Tiber river were, at tropically high temperatures, extremely wearing. The fast push of the American 5th Army could only be stopped in the middle of June near Chiusi, which had primarily been stubbornly defended by the Fallschirm-Panzer-Regiment HG for some time, and at the lake of Trasimeno. Then, when the division had reached the south shore of the Arno, it was pulled out on June 15, 1944, assembled between Bologna and Ferrara, and front. There, a catastrophic situation had developed for all of the eastern front, the Russians had broken through the Heeresgruppe Mitte (central army group). To stuff the gaping hole there, a number of combat ready divisions had to be withdrawn from other theaters of war. Among them was also the Fallschirm-Panzer-Division HG.

Thus, the division moved out on July 24, 1944 for a three-day journey through the Brenner pass, via Innsbruck, Munich, Regensburg, Eger, Aussig, Dresden, Goerlitz, Lauban, Breslau, Ostrowo, Lodz, Skiernevice in the direction of Warsaw.

Studying the maps.

Various patches were worn with the black Panzer jackets: black with white or pink borders, white without border with death's head, or, instead of the patch, only the death's head. A common regulation was obviously not known. The troops accepted it, they had other concerns. Left, Major Hahm, commander of the II./Fallschirm-Panzer-Regiment HG.

Allied bomber units knocked out all supply routes.

Vehicles were easy prey for the fighter bombers constantly circling in the sky.

Sending a quick greeting home.

Changing a track roller.

On historical ground: the Via Appia Antica on which Caesar's legionnaires already marched.

The further east we came, the more new dangers lurked: through guerilla activity which became more noticeable, particularly at night, and through epidemics which also threatened the troops.

Typhoid. Any stop by non-resident troops in Tolfa is forbidden. Tolfa is a typhoid area.

Warning! Guerilla danger. Drive in convoys at night.

Camouflage was everything, in particular also for the supply troops. During the day any vehicle was attacked by fighter bombers. Thus, movement took place only at night. During the day the vehicles were carefully camouflaged to keep them out of sight of the pilots. Even from a short distance, as on this photo, the truck is hardly visible.

(Only those parts of the division in action in Italy)

Officers
Noncoms
Ranks
Foreign volunteers
Total
Required
Available
Shortfall

Reinforcements arrived during the month of May:

17 Officers
616 Noncoms and ranks
Total 633 men
Losses from 1 to 31 May 1944
Officers
Noncoms and ranks
Total
Fallen
Wounded
Missing
Ill
Others
Total

The high losses occurred particularly during the last days of the month of May when the division had to move from the resupply area in Tuscany in day marches to the combat area Valmontone-Rome, and in the fighting here during which the units were greatly dispersed and some units totally scattered.

Die Personallage der Fallschirm-Panzer-Division HG am 1. Juni 1944

(Nur Teile der Division, die in Italien eingesetzt sind)

	Offi-ziere	Unter-offiziere	Mann-schaften	Hiwis	Gesamt
Soll	658	4 726	16 272	77	21 733
Ist	568	3 353	15 599	38	19 558
Fehl	90	1 373	673	39	2 175

Im Monat Mai eingetroffener Ersatz: 17 Offiziere, 616 Unteroffiziere und Mannschaften, zusammen 633 Mann.

Verluste vom 1. bis 31. Mai 1944

	Offi-ziere	Unteroffiziere und Mannschaften	Gesamt
Gefallen	11	255	266
Verwundet	19	402	421
Vermißt	11	952	963
Krank	10	629	639
Sonstige	11	247	258
Gesamt	62	2 485	2 547

Die hohen Verluste sind besonders in den letzten Tagen des Monats Mai entstanden, als die Division aus dem Auffrischungsraum in der Toskana in den Kampfraum Valmontone-Rom im Tagesmarsch verlegen mußte, und in den Kämpfen hier, in denen die Verbände stark zersplittert und einige Einheiten völlig versprengt wurden.

To make soldiers available for duty at the front, foreign volunteers were moved into soldiers' posts, in particular into supply and repair units. Here seen is a group of Italian volunteers who were used in the administration company as loaders, transport escorts and for duties in the depots.

The weapons of the Fallschirm-Panzer-Division HG on June 1, 1944.

Die Bewaffnung der Fallschirm-Panzer-Division HG
am 1. Juni 1944

Waffenart	Soll	Ist
Karabiner und Gewehre 98	15 134	15 385
Maschinenpistolen	2 185	1 871
Leichte Maschinengewehre	1 354	906
Schwere Maschinengewehre	86	73
Maschinengewehre in Panzern	208	160
Granatwerfer	70	51
Schwere Panzerbüchsen 41	241	7
2 cm-Flak 38	91	70
2 cm-Flak-Vierlinge	42	42
3,7 cm-Flak 37	12	8
8,8 cm-Flak 36 und 37	36	35
Leichte Infanterie-Geschütze	28	16
Schwere Infanterie-Geschütze	14	14
Leichte Feldhaubitzen 18	24	21
Schwere Feldhaubitzen 18	16	16
10 cm-Kanonen 18	8	7
Schwere Feldhaubitzen Hummel	6	-
7,5 cm-Pak 40	58	35
15 cm-Nebelwerfer	9	12
Schwere Wurfgeräte 41	12	9
3,7 cm-Pak (mot. und SPW)	4	4
2 cm-Kampfwagen-Kanonen Spähw.	60	8
7,5 cm-Kanonen 37 (Sf.)	30	14
Ofenrohre	60	144
Flammenwerfer	80	69

Category of weapon
Required
Available
Carbines and rifles 98
Machine pistols
Light machine-guns
Machine-guns in Panzers
Mortars
Heavy anti-tank rifles 41
2 cm. flak 38
2 cm. Flak four-barrelled
3.7 cm. Flak 37
8.8 cm. Flak 36 and 37
Light infantry guns
Heavy infantry guns
Light field-howitzers 18
Heavy field-howitzers 18
10 cm. canons 18
Heavy field-howitzers Hummel (bumble-bee)
7.5 cm. Pak 40 (anti-tank gun)
15 cm rocket launcher
Heavy mortars 41
3.7 cm. Pak (motorized and on armored cars)
2 cm armored car canons
7.5 cm. canons 37 (self-propelled)
Bazookas
Flame throwers

The vehicle situation of the Fallschirm-Panzer-Division HG on June 1, 1944.

Die Fahrzeuglage der Fallschirm-Panzer-Division HG am 1. Juni 1944 (Nur Teile der Division, die in Italien eingesetzt sind)					
Fahrzeug	Soll	Ist	Davon: einsatz- bereit	zur In- stand- setzung	Fehl
Gepanzerte Fahrzeuge					
Sturmgeschütze	31	16	8	8	15
Panzer III	19	13	2	11	6
Panzer IV	98	56	18	38	42
Schützenpanzerwagen	366	306	280	26	60
Pak Selbstfahrlafette	28	16	7	9	12
Kraftfahrzeuge					
Krafträder	864	716	645	71	148
Personenkraftwagen	1 100	868	662	206	232
Lastkraftwagen	2 700	1 318	1 027	291	1 382
Tonnage (t)	6 919	3 924	2 918	1 006	2 995
Zugkraftwagen	322	148	125	23	174

Die Soll-Zahlen sind mit, die Ist-Zahlen dagegen ohne die Fahrzeuge der in Holland zur Umrüstung auf Panzer V "Panther" befindlichen I. Abteilung des Panzer-Regiments

Vehicle
Required
Available
Ready for action
Under repair
Shortfall
Armored vehicles
Assault guns
Panzer III
Panzer IV
Armored personnel carriers
Self-propelled Pak
Vehicles
Motorcycles
Cars
Trucks
Tonnage
Tractors

The 'required' figure are with, the 'available' figure are without the vehicles of the I. Abteilung of the Panzer-Regiment which is now being re-equipped with Panzerv "Panthers" in Holland.

The German soldiers fallen north of Rome were moved after the war to the German war cemetery Futa-Pass. Here now rest 30,635 dead, a large portion of them "Unknown" which can be explained by the atrocious fighting tactics of the Italian guerrillas. In the photo the memorial designed by Professor Oesterlen.

Meldung vom _1. Juli_ 1944   Verband: Pz.Gr.Div.H.G.   Unterstellun sverhältnis: LXXVI.Pz.Korps

1. Personelle Lage am Stichtag der Meldung:

a) Personal:

	Soll	Fehl
Offiziere	655	133
Uffz.	5 162	1 877
Mannsch.	15 368	2 220
Hiwi	972	852
insgesamt	22 157	5 112

c) in der Berichtszeit eingetroffener Ersatz:

	Ersatz	Genesene
Offiziere	18	4
Uffz. und Mannsch.	247	171

b) Verluste und sonstige Abgänge in der Berichtszeit vom 1.6.44 bis 30.6.44

	tot	verw.	verm.	krank	sonst.
Offiziere	19	45	26	14	17
Uffz. und Mannsch.	411	1101	920	841	245
insgesamt	430	1144	946	855	262

d) über 1 Jahr nicht beurlaubt:

insgesamt	1511	Köpfe 8	d. Iststärke
davon:	12-18 Monate	19-24 Monate	über 24 Monate
	1505	6	-

Platzkarten im Berichtsmonat zugewiesen: 2800

2. Materielle Lage:

		Gepanzerte Fahrzeuge							Kraftfahrzeuge				
		Stu. Gesch.	II	IV	V	(...)	Pak SF		Kräder			Pkw	
									Ketten	ang tr Bew	sonst.	gel.	O
Soll (Zahlen)		31	3	98	3	12	341	28	132	306	425	957	143
einsatzbereit	zahlenm.		2	12	-	9	118		-	155	435	334	273
	in % des Solls		66,7	12,2	-	75	34,5		-	61,3	102,3	34,7	14,3
in kurzfristiger Instandsetzung (bis 3 Wochen)	zahlenm.		4	17	-	1	27		-	71	122	114	92
	in % des Solls		133,3	17,3	-	8,3	7,9		-	23,1	28,5	11,9	64,3

		noch Kraftfahrzeuge						Waffen				
		Lkw				Ketten-Fahrzeuge		s Pak 7,5 9	art. Gesch	MG ( )	sonstige Waffen	
		Maultier	gel.	O	Tonnage	Zgkw	RSO					
Soll (Zahlen)		185	1057	1457	6919	168	154	30 21	54	1440		
einsatzbereit	zahlenm.	17	253	724	2952	48	47	13 4	34	759 (470)		
	in % des Solls	9,2	13,3	49,7	42,9	28,8	30,6	43 19	63	53		
in kurzfristiger Instandsetzung (bis 3 Wochen)	zahlenm.	3	84	247	1044	10	12	4 3	6	(215)		
	in % des Solls	1,6	7,9	13,5	15	6	7,8	13 14	11	4		

o) **Werfer**

3. Pferdefehlstellen

*) Zgkw. mit 1 5 t, **) Zgkw. mit 8 18 t
(davon MG. 42

Kurzes Werturteil des Kommandeurs:

a) **Ausbildungsstand: gut.**

b) **Stimmung der Truppe:** gefestigt und **gut.**

c) **Besondere Schwierigkeiten:** ungewöhnliche schlechte Ersatzteillage, insbesondere bei Panzerersatzteilen und Reifen. Bei dem an sich schon stark abgesunkenen Kfz.-Ist führt die hohe Zahl nicht einsatzbereiter Kfz. zu Versorgungsschwierigkeiten (insbesondere Muni.-Transport). Ein Teil der Truppe kann nur durch zweimaliges Fahren verlastet bewegt werden. Wegen des grossen Mangels an Zugmaschinen können 2 Artl.-Battr. und 3 s.Flak-Battr. nicht zum Einsatz gebracht werden.

d) **Grad der Beweglichkeit in % des Solls:** 45 - 50 %.

e) **Kampfwert und Verwendungsmöglichkeit:**
Zu jeder Angriffsaufgabe geeignet.

Kurze Stellungnahme der vorgesetzten Dienststelle:

Durch Auskämmen der Trosse und Ersatzzuführung ist Gefechtsstärke erheblich gehoben worden.

Division hat sich voll bewährt.

General der Panzertruppen und
Kommandierender General des
LXXVI.Pz.-Korps.

Short assessment by the commander: a) state of training: good. b) morale of the troops: firm and good. c) particular difficulties: unusually bad replacement parts situation, in particular with tank parts and tires. With the already heavily reduced number of available vehicles, the high number of not operational vehicles leads to supply problems (especially ammunition transport). A portion of the troops with equipmenoftractos arilley bateies and 3 Flak batteries cannot be brought into action. d) level of mobility in % or required: 45-50%. e) combat readiness and possible use: fit for any attack assignment.

Short comment of the superior department: By combing baggage trains and additional replacements the combat readiness has been greatly increased. The division has fully stood the test.

All Panzer divisions had to submit to the inspector general of the Panzer forces monthly situation reports through channels. This was meant to provide the leadership with a clear picture of the materiel conditions and the morale of the troops.

At the end of July 1944, the Division rolled to the eastern front with Warsaw as the destination.

The Fighting at the Weichsel Bend

"Announced" by the partisans, the transport trains arriving in Warsaw from Italy were greeted, during the unloading, by bombs and shelling. The battalions and detachments were thrown into the fighting as they arrived-and not in the tactically correct sequence. During the first days they fought initially south-east of Warsaw. Subsequently, the Fallschirm-Panzer-Division HG was successful, together with the units of the army, in encircling the III. Soviet tank corps which had broken through near Wolomin-Radczymin. On August 3, 1944, it was crushingly defeated and German fighter planes, as was rare in these days, provided excellent assistance. As Generaloberst Model stated in the orders of the day for the Heeresgruppe Mitte, it was thanks to the immediate and courageous engagement of the Fallschirm-Panzer-Division HG that Warsaw could be held one more time, and the Russians were pushed back from the east bank of the Weichsel river.

When the Soviets crossed the Wechsel at the mouth of the Pilica river between Warka and Magnuszew during the early days of August 1944 and formed a bridge-head of ten kilometers wide and three to five kilometers deep, the Fallschirm-Panzer-Division HG, which had become the "fire brigade" of the 9. Armee, was brought into action on August 8, 1944 within the XXXXI. Panzerkorps and occasionally within the VIII. Armeekorps. Throughout the month of August heavy and varied fighting took place here. Still, the widening of the bridge-head to the operational breakthrough towards Radom could be prevented. The Fallschirm-Panzer-Division HG had a major share in the defensive successes.

In September and October the Division fought in the area of the 2. Armee (Generaloberst Weiss) east of the Weichsel between Warsaw and Modlin in the sector of the IV.SS-Panzerkorps. The report of the armed forces mentioned the equally successful engagement of the Division at the eastern front three times: on August 3, 1944, that "a company of the Fallschirm-Panzer-Division HG under the leadership of Hauptmann Bellinger destroyed 36 tanks within 24 hours during heavy tank duels"; on August 5, 1944, that "the Fallschirm-Panzer-Division HG continued its attacks east of Warsaw on the encircled but doggedly resisting Bolsheviks and enemy relief attacks have faltered"; and on September 24, 1944, that, in the Warsaw area, a task force made up partly from units of the Fallschirm-Panzer-Division HG "distinguished itself through singular determination and initiative".

Sturmgeschutz III (assault gun) with its crew.

Fsch.-Panzerdivision Hermann Göring   Div.Gef.St., den 25.7.1944
- Ia - Br. B. Nr. 1372/44 geh. -

G E H E I M

D i v i s i o n s b e f e h l   Nr. 2/44

für das Eintreffen im Versammlungsraum der Division.

1.) Feind stößt mit Panzerkräften aus dem Raum Lublin und ostwärts in Richtung auf Warschau - Jadow vor.

Brückenkopf ostw. Warschau wird z.Zt. durch eigene schwache Sicherungskräfte geschützt. Vorstoß von Feindkräften über die Weichsel südlich Warschau und in nordwestlicher Richtung ist möglich.

2.) Fsch.Pz.Div.H.G. versammelt sich mit Kampfteilen im Raum Rembertow, mit Versorgungsteilen im Raum Warschau (ausschl.) - Modlin zum voraussichtlichen Einsatz ostw. der Weichsel in südostw. Richtung.

3.) Nach Eintreffen des Transportzuges auf dem Auslade-Bhf. (im allgemeinen Warschau-West oder Praga Vorstadt) ist die Auslaung unverzüglich durchzuführen, die Truppe ist hierbei von der Auslaestrecke abzusetzen, auseinanderzuziehen und zu tarnen (Fliegerdeckung!).
Jeder Transportführer meldet sein Eintreffen unverzüglich persönlich oder fernmündlich beim Auslade-Offz.der Div. auf Bhf. Warschau-West mit Angabe von Fahrt-Nr. und Truppenteil.

4.) Nach beendeter Ausladung ist in den Unterkunftsraum anzumarschieren. Unterbringung der Truppenteile gem. Skizze. Der Warschau-Westbhf. zum Div.-Gef.Stand ist ausgeschildert. Zur Erkundung des Unterkunftsraumes ist nach Eintreffen auf dem Auslade-Bhf ein Vorkdo. vorauszusenden, das sich grundsätzlich zunächst beim Div.-Stab zu melden hat. Unterlagen über Kopfstärke, Kfz. und Waffenzahl sowie Versorgungslage (Betriebsstoff, Verpflegung, Munition) sind mitzubringen.

5.) Sicherungsmaßnahmen:
Im Unterkunftsraum der Truppenteile hat sich jede Einheit nach näherer Weisung ihres Truppenteiles bzw. selbständig so zu gliedern, dass sofortige Abwehr feindl. Überfällen (Angriff von der Erde, Bandentätigkeit, Fallschirmjäger) gewährleistet ist. (Igel!) Alle Straßen und Zufahrtswege von Osten und Südosten sind zu sichern, verfügbare schwere Waffen sind einzusetzen. Luftlage erfordert sorgfältigste Tarnung (wie in Italien!). Kfz. sind einzugraben und sofort Deckungslöcher für alle Mannschaften zu schaffen. Alle verfügbaren Fliegerabwehrwaffen (auch MG) sind einzusetzen.
Schnelle Versammlung zum Abmarsch in südlicher bzw. ostw. Richtung ist vorzubereiten. Die Truppe muß spätestens nach 2 Stunden abmarsch- und voll einsatzbereit sein.

6.) Einsatz der Truppe erfolgt ausschließlich auf Befehl der Div. Anforderungen anderer Dienststellen sind unter Verweisung auf die Div. abzulehnen. Das erfolgte Eintreffen im Unterbringungsraum ist durch Unteroffiziers-Melder dem vorgesetzten Truppenteil bzw. der Div. unmittelbar zu melden. Der Melder verbleibt dann zunächst als Befehlsüberbringer beim Div.-Stab.

7.) Auf anliegendes Merkblatt wird hingewiesen. Die Truppe ist entsprechend zu belehren.

Verteiler: Für das Divisionskommando
Bis Transp.-   Der erste Generalstabsoffizier
führer
- 1 - Anlage   [signature]

Secret. Division order 2/44 for the arrival at the assembly area of the Division. 1) Enemy advances with tank forces from the area of Lublin and east towards Warsaw-Jadow. Bridge-head east of Warsaw is at present being secured by own weak forces. Advance of enemy forces across the Weichsel south of Warsaw and in north-westerly direction is possible. 2) Fsch.Pz.Div. HG assembles its fighting forces in Rembertow area, the support units in the Warsaw-Modlin area for probable action east of the Weichsel in a southeast direction. 3) After the arrival of the transport train at the station (either Warsaw-west or Praga suburb), unloading is to be carried out immediately, the troops are to be removed from the unloading area, spread and camouflaged (air-raid cover). Each transport leader reports his arrival immediately in person or by telephone to the unloading officer of the Division at the Warsaw-west station, including transport number and unit. 4) After unloading the troops are to march to the billet area as shown on the sketch. The route from Warsaw-West station to divisional headquarters is marked. After arrival at the station an advance party is to be sent to scout the billet area. It must first report to the divisional staff with documentation on numbers of men, weapons, vehicles as well as supplies (Fuel, rations, ammunition) 5) Security measures: In the billet area each unit must so deploy, as directed in detail or independently, that immediate defense against enemy attacks (ground attack, guerilla activities, paratroopers) is guaranteed. All roads and approaches from the east and southeast are to be secured, available heavy weapons to be deployed. Air situation requires most careful camouflage (as in Italy!). Vehicles are to be dug-in, fox-holes for all men are to be dug. All available anti-aircraft weapons (including machine-guns) are to be deployed. Preparation for immediate departure into southerly or easterly direction is to be made. The troops must be ready to move and for action within 2 hours of marching order. 6) Units will deploy exclusively on order from the Division. Requests from other headquarters' are to be refused with referral to the Division. The arrival in the billet area must be reported immediately through noncom dispatch to the superior unit or the Division. The noncom will initially remain at divisional staff transmission of orders. 7) Attention is drawn to the attached memorandum. The troops are to be instructed accordingly.

The Fallschirm-Panzer-Division HG played a decisive part in the successful defensive battle before Warsaw.

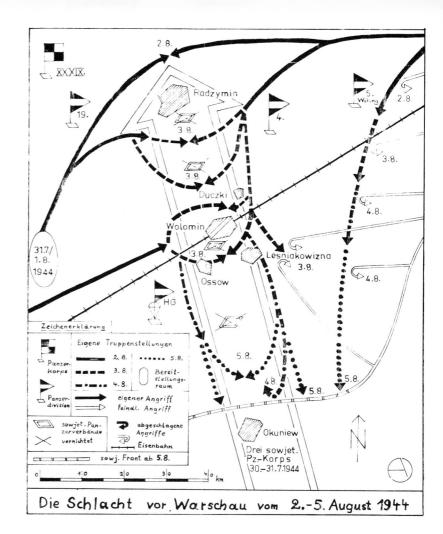

A Soviet tank seen through the optics of an assault gun.

A neighboring tank is hit during the tank battle near Porgorzel.

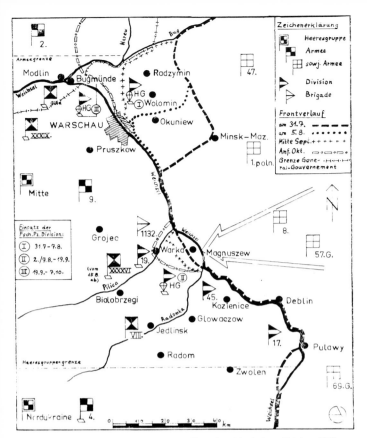

Zeichenerklärung

	Heeresgruppe
	Armee
	sowj. Armee
	Division
	Brigade

Frontverlauf

am 31.7. — — — —
am 5.8. • • • • •
Mitte Sept. + + + + +
Anf. Okt. ▭▭▭▭▭
Grenze General-Gouvernement

Einsatz der Fsch.Pz.Division:

I 31.7 – 7.8.
II 2./7.8. – 19.9.
III 19.9. – 7.10.

Brückenkopf Warka – Magnuszew von Ende Juli bis Anfang Oktober 1944

Points of major action by the Fallschirm-Panzer-Division HG and the Fallschirm-panzerkorps HG, created here, was the fighting east of the Weichsel between Warsaw and Modlin and at the Weichsel bridge-head Warka-Magnuszew.

With the uncertain enemy situation east of Warsaw, reconnaissance and immediate relay of results were particularly important to ensure cooperation with the grenadiers.

Because of the air activity, brisk also on the eastern front, good camouflage of the Panzers was a requirement.

128

The supreme commander of the Armee in discussion with Generalmajor Schmalz, just appointed commanding general, and Oberstleutnant i.G. von Baer, newly appointed chief of the general staff of the only Panzerkorps of the Luftwaffe.

The anti-aircraft guns had set up as an anti-tank block and fired, when the situation required it, from the carriage, taking direct aim.

Doggedly, the Pak defended against the approaching tank packs east of Warsaw.

The Replacement and Training Regiment HG

Each combat unit had, in the home land or later also in the occupied territories, a replacement unit which trained its replacement personnel. At the mobilization of the RGG, the I. Abteilung RGG established the replacement unit RGG in Berlin-Reinickendorf which consisted of the training staff, four placement batteries and a convalescent battery. This replacement unit was moved after them Western Campaign to Utrecht, Holland.

With the expansion of the RHG to the Brigade HG and to the Division HG the replacement unit was enlarged to the replacement regiment HG-later named units in Utrecht, Amersfort and Hilversum. The IV. unit (driver training) was at times stationed in Velten near Berlin.

For most of the young volunteers the Replacement Regiment HG was the first unit they came to know in their lives and where they received their training and, many of them, additional special training during later courses. For almost all the replacement unit was, after their wounds had healed, the way station from the hospital back to "their bunch".

When they were taken on strength, each recruit received as personal identification the pay-book and the identification tag. The latter had to always be worn on a string around the neck. In case of death the tag was broken at the burial. The top part remained for later identification (e.g. when the body was moved to another grave) with the dead, the lower part had to be attached to the casualty report which contained the location of the grave.

The replacement unit of the RGG was stationed in Utrecht since 1940: commander was then Major Grauert (left in the photo).

Soon after being taken on, the swearing-in took place as here in May 1941 at the Krumhout barracks in Utrecht. A soldier from each group took the oath in front, as he touched the pilot's sword of his officer.

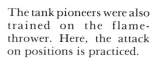

The tank pioneers were also trained on the flame-thrower. Here, the attack on positions is practiced.

The main emphasis during basic training was the infantry training of the group, here on the light machine-gun.

Training in the Krumhout barracks in Utrecht.

With the establishment of the Division HG, the replacement unit was enlarged to the Replacement and Training Regiment HG whose commander, for a long time, was Major Schulz (in the photo front group right). In Holland, the Regiment reported to the commander of the armed forces in the Netherlands, General der Flieger (general of the pilots) Christiansen, who is here inspecting the Regiment (front group center).

Training on the "Achtacht" (8.8 cm gun).

IM NAMEN
DES OBERBEFEHLSHABERS
DER LUFTWAFFE

verleihe ich dem

Lw.-Oberhelfer Walter O q u e k a

2./schw. Flakabteilung 394

DAS KAMPFABZEICHEN
DER FLAKARTILLERIE

der Kommandeur der 4. Flak-Division

STABSQUARTIER,
den 20.2.44
NR: 3146

Generalmajor

'In the name of the Supreme Commander of the Luftwaffe I award the Lw.-Oberhelfer (air force helper) Walter Oqueka the fighting badge of the Flak artillery. The Commander of the 4 Flak-Division at staff headquarters, 2/20 1944.'

The students of high schools were often trained on the Flak and the classes deployed as units as helpers with the Luftwaffe after the motto: 'Learn during the day-fight during the night.' After a lengthy 'tour of duty' already promoted, and awarded, for succesful participation in downing aircraft, with the Flak fighting badge, Walter Oqueka volunteered, when his age group was drafted, for the Fallschirm-Panzer-Division HG in the hope of joining the Flak-Panzer-Division HG in the hope of joining the Flak-Regiment HG as a fully trained and 'seasoned' gunner. He was wrong. Despite his protests he was made a Panzergrenadier.

The Guard Unit in Karinhall

In 1935, the Reichsminister for Aviation and Supreme Commander of the Luftwaffe, Göring, had chosen the woodland property Karinhall, located at the Grossen Dolln lake in the center of the Schorfheide (moor) some 50 kilometers north of Berlin, to be his stately official residence. Guarding it was the responsibility of the RGG.

During the years of peace the Wach-Batallion/RGG (guard battalion) provided the frequently changing guard at strength of a platoon, later a full guard company. During the war a guard company and a 2 cm Flak battery were constantly deployed there. In addition, a heavy Flak unit was stationed there to secure Karinhall, it was, however, not part of the RGG. To deceive the enemy air reconnaissance, a fake installation, 'dummy Karinhall', had been built some ten kilometers from Karinhall. Karinhall was spared any enemy action until the end of the war.

The soldiers who served in Karinhall were subject to stringent regulations for security reasons. They were compensated by being able to witness many high level political events from the sidelines. Also, the relationship of the Göring family and their house personnel to the guard soldiers was relaxed and friendly, so that duty there was generally enjoyed.

When Berlin was already encircled in April 20, 1945, Karinhall directly threatened by Russian troops, the Göring family left their domicile. Very early on April 20, 1945, Karinhall, which had already been prepared for this by pioneers, was blown up. Göring, himself, was supposed to have activated the ignition. Today, one can hardly find the overgrown remnants of the walls.

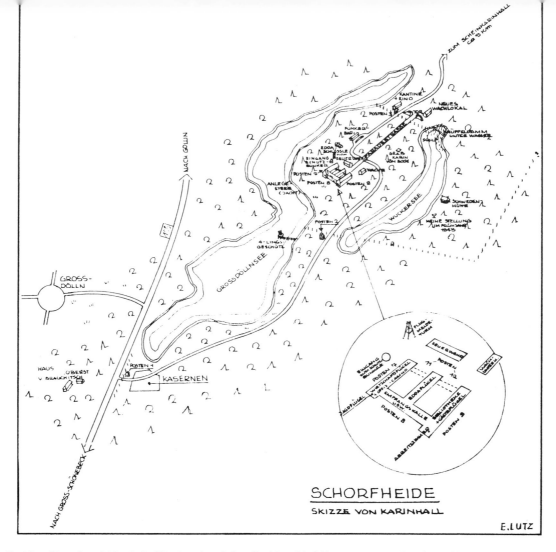

'Schorfheide. Sketch of Karinhall'. In the Schorfheide, 50 kilometers north of Berlin, the Reichsminister for Aviation and Supreme Commander of the Luftwaffe had established his stately official residence in the woodland property Karinhall.

The entrance gate to Karinhall, start of the security zone.

Guard at the Gross-Dolln lake.
Across the woodland property
Karinhall.

Fire-place corner in the reception
hall.

The Fallschirmpanzerkorps HG
within the Fallschirm-Armee 1.

(Solid line: administrative authority;
broken line: operational authority).

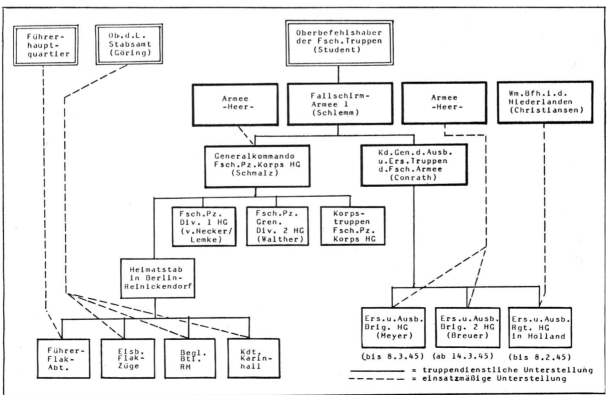

The Establishment of the Fallschirmpanzerkorps HG

Despite the burden on leadership and the troops caused by heavy fighting, the enlargement of the Fallschirm-Panzer-Division HG to the Fallschirmpanzerkorps HG took place in October 1944. This expansion occured partly through splitting of existing units, partly through adding of whole units which had been raised at the replacement formations or the training areas, or through the inclusion and re-naming of existing units.

In its final format the Fallschirmpanzerkorps HG was structured as follows:

I. Generalkommando
   — Kommandierender General: Generalleutnant Schmalz —

II. Korpstruppen:
   Korps-Sturm-Bataillon HG
   Korps-Panzerjäger-Abteilung HG
   Korps-Pionier-Bataillon HG
   Korps-Nachrichten-Abteilung HG
   Fallschirm-Flak-Regiment HG mit vier Abteilungen
    I. Nachschub-Abteilung HG
   II. Instandsetzungs-Abteilung HG
   Verwaltungs-Bataillon HG
   Korps-Sanitäts-Abteilung HG
   Korps-Feldpostamt HG

III. Fallschirm-Panzer-Division 1 HG
   — Divisionskommandeur: Generalmajor von Necker,
     ab Februar 1945 Generalmajor Lemke —
   Divisionsstab
   Fallschirm-Panzer-Regiment HG mit zwei Abteilungen
   Fallschirm-Panzergrenadier-Regiment 1 HG mit zwei Bataillonen
   Fallschirm-Panzergrenadier-Regiment 2 HG mit zwei Bataillonen
   Fallschirm-Panzerfüsilier-Bataillon 1 HG
   Fallschirm-Panzeraufklärungs-Abteilung 1 HG
   Fallschirm-Panzerpionier-Bataillon 1 HG
   Fallschirm-Panzerartillerie-Regiment 1 HG mit drei Abteilungen
   Fallschirm-Panzernachrichten-Abteilung 1 HG
   Feldersatz-Bataillon 1 HG
   Sanitäts-Abteilung 1 HG
   Feldpostamt 1 HG

IV. Fallschirm-Panzergrenadier-Division 2 HG
   — Divisionskommandeur: Generalmajor Walther —
   Divisionsstab
   Fallschirm-Sturmgeschütz-Abteilung HG
   Fallschirm-Panzergrenadier-Regiment 3 HG mit drei Bataillonen
   Fallschirm-Panzergrenadier-Regiment 4 HG mit drei Bataillonen
   Fallschirm-Panzerfüsilier-Bataillon 2 HG
   Fallschirm-Panzeraufklärungs-Abteilung 2 HG
   Fallschirm-Panzerpionier-Bataillon 2 HG
   Fallschirm-Panzerartillerie-Regiment 2 HG mit drei Abteilungen
   Fallschirm-Panzernachrichten-Abteilung 2 HG
   Feldersatz-Bataillon 2 HG
   Sanitäts-Abteilung 2 HG
   Feldpostamt 2 HG

V. Sonstige Verbände und Dienststellen:
  Führer-Flak-Abteilung (nur noch in der ersten Zeit)
  Begleit-Bataillon Reichsmarschall
  Fallschirm-Panzer-Ersatz- und Ausbildungs-Brigade HG mit den
    Regimentern 1 und 2 (bis März 1945)
  — Brigadekommandeur: Oberst Meyer —
  Fallschirm-Panzer-Ersatz- und Ausbildungs-Brigade 2 HG mit den
    Regimentern 3 und 4 (ab März 1945)
  — Brigadekommandeur: Oberst Breuer —
  Ersatz- und Ausbildungs-Regiment HG in Holland
  Heimatstab Fallschirmpanzerkorps HG in Berlin-Reinickendorf
  Leichtkranken-Lazarette (Erholungsheime) in Oberau und Reit i. W.

The establishment dragged on over a longer time period, caused by the extremely noticeable shortage of personnel, weapons, and especially vehicles in the sixth war year, and planned strength and reserves were not always achieved. The reduced mobility of the units, caused by the shortage of fuel, lead often to the substitution of customary horse drawn carts. According to the Landser humor, the Fallschirmpanzerkorps became the Fallschirmpanjekorps! (Russian for horse)

The commander of the Fallschirm-Panzergrenadier-Division HG: Generalmajor Walther.

The commanding general of the Fallschirmpanzerkorps HG: Generalmajor Schmalz. On January 30, 1945 he was promoted to Generalleutnant.

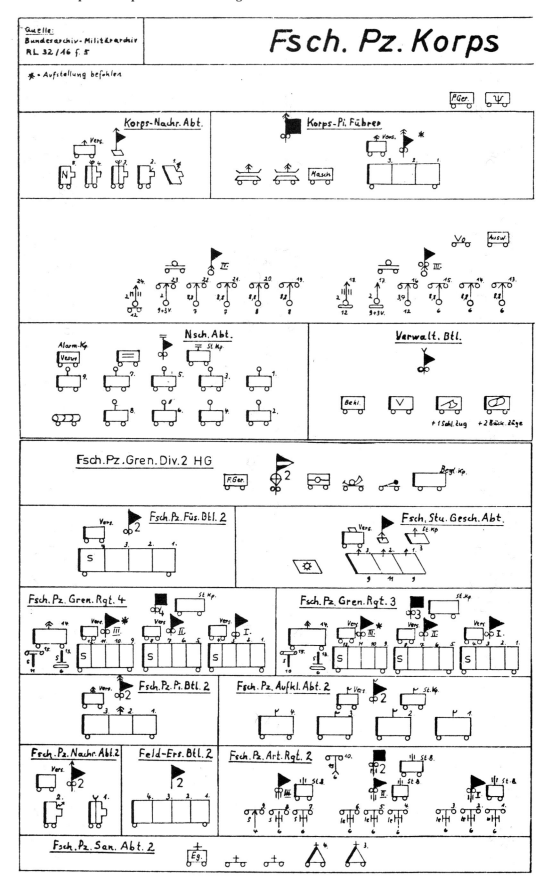

Hermann Göring

Begl.Kp.

Korps-Art.Führer

Korps-Pz.Jäger-Abt.
Vers.
St.Kp.
IV IV IX
6   6   6

Korps-Sturm-Btl.
Vers.
4 3 2 1
S

Fsch.Flak-Rgt.

II.
12 11 10 9 8 7
2 2 3,7 8,8 8,8 8,8
12 9+3V 13 6 6 6

I.
6 5 4 3 2 1
2 2 3,7 8,8 8,8 8,8
10 9+3V 12 6 6 6

Korps-San.Abt.

+ Zahn.St.

Inst.Abt.

6. G
Flak.Sd.
7.
Flak3nst.
8.

4.
X Wffm.

1. + Pz.Werkst.Zg.

2.

Ers.

Fsch.Pz.Div.1 HG
F.Ger. 1 Begl.Kp.

Fsch.Pz.Füs.Btl.1
Vers.
4 3 2 1
S

Fsch.Pz.Rgt.
Vers. St.Kp.
II.
IV IV IV IV
6   6   11 12
St.Kp.
Vers. I.
4 3 2 1
V V V V
9 10 10 11

Fsch.Pz.Gren.Rgt.2
2 St.Kp.
10. Vers. II. Vers. I.
11 9 8 7 6 5 3 2 1
S S

Fsch.Pz.Gren.Rgt.1
1 St.Kp.
10. Vers. II. Vers. I.
11 9 8 7 6 5 3 2 1
S S

Fsch.Pz.Pi.Btl.1
Vers. 1

Fsch.Pz.Aufkl.Abt.1
Vers. St.Kp.
4 3 2 1

Fsch.Pz.Nachr.Abt.1
Vers. 1
2 1

Feld-Ers.Btl.1
1
4 3 2 1

Fsch.Pz.Art.Rgt.1
1 St.B.
III. St.B.
II. St.B.
I. St.B.
9 8 7 6 5 4 3 2 1

Fsch.Pz.San.Abt.1
2. 1.

139

er Reichsmarschall des Großdeutschen Reiches H.Qu., den 24.9.1944
und Oberbefehlshaber der Luftwaffe
11b12.16 Nr.13266/44 g.Kdos.(Genst.Gen.Qu.2.Abt.,IIB)

r.: Aufstellung Fsch.Pz.Korps Hermann Göring. 95 Ausfertigungen
g.: -2- Ausfertigung.

I. 3986/44

1.) Mit sofortiger Wirkung wird
aufgestellt das

Fsch.Pz.Korps Hermann Göring.

2.) Aufstellungsraum: Modlin.
Aufstellungsorte im einzelnen sind durch Fsch.A.O.K. fest-
zulegen.

3.) Gliederung gemäss Anlage 1.

4.) Durchführung durch Fallschirm A.O.K.

II.
Zuführung von Einheiten:

Dem Fsch.Pz.Korps H.G. werden mit sofortiger Wirkung zugeführt:

1.) Begleit-Rgt. H.G. mit
Stab/Begl.Rgt.H.G.
I. (Gren.)/Begl.Rgt.H.G. (ohne 2.Komp.)
II.(Flak)/Begl.Rgt.H.G. (ohne 12.Batterie)

2.) Fsch.Jg.Rgt.16 mit
Stab und Stabskomp.
Pionier-Komp.
s.I.G.-Komp.
Pz.Jg.Komp.
I./Fsch.Jg.Rgt.16
II./ " " " 16
III./" " " 16

3.) Nachrichtentruppen
2 Pz.Nachr.Abt.
1 Korps-Fernspr.Komp.

'The Reichsmarschall of the Greater German Reich and Supreme Commander of the Luftwaffe. Headquarters, 9/24/1944. Re: Establishment of the Fsch.Pz.Korps Hermann Göring. I. 1.) To be established with immediate effect: the Fsch.Pz.Korps Hermann Göring. 2.) Assembly area: Modlin. Detailed locations to be determined by the Fsch.A.O.K. 3.) Organizations as per attachment 1. 4.) Implementation by Fallschirm A.O.K. II. Attachment of units: To be attached to the Fsch.Pz.Korps HG with immediate effect: 1.) Begleit-Rgt. HG with Stab/Begl.Rgt. HG, I. (Gren.)/Begl.Rgt. HG (without 2.Komp.), II. (Flak)/Begl.Rgt. HG (without 12. Batterie); 2.) Fsch.Jg .Rgt.16 with staff and staff comp., Pionier-Komp., s.I.G.-Komp., Pz.Jg.Komp. I./Fsch.Jg.Rgt.16, II./Fsch. Jg.Rgt.16, III./Fsch.Jg.Rgt.16, 3.) Communications units: 2 Pz.Nachr. Abt., 1 Korps-Fernspr.Komp.'

The birth certificate of the Fallschirmpanzerkorps HG. The first page of the 11-page establishment order.

The commanders of the Fallschirm-Panzer-Division 1 HG: (left) Generalmajor von Necker, and his successor Generalmajor Lemke.

The Escort Regiment HG and the Escort Battalion Reichsmarschall

The Wach-(guard) Regiment HG was dissolved as such in April 1944. From it was created the Begleit-(escort) Regiment HG in a completely new composition. The I. Batallion was a Panzergrenadier-Batallion with six companies and the cavalry platoon, the II. Abteilung was a Flak-Abteilung. Additionally, the Flak-Abteilung 194 was taken over which also included the Eisenbahn-(railroad) Flak-Begleit-(escort) Batterie in which the railroad escort trains were contained.

In July 1944 the Begleit-Regiment HG was moved to East Prussia as a combat unit. It was responsible for the security of the Reichsmarschall, the Reichjagerhof (hunting lodge) in Rominten, and the headquarters of the supreme commander of the Luftwaffe. Apart from units of the Begleit-Regiment which were charged with specific duties, the Regiment was put into action in the East Prussian border area and in Lithuania (here sometimes as a part of the combat group Schirmer (Fallschirm-Jager-Regiment 16)) from July to September 1944.

In the process of the establishment of the Fallschirmpanzerkorps HG, the Begleit-Regiment HG was dissolved in October 1944. It formed the regimental staff and the II. Batallion of the Fallschirm-Panzergrenadier-Regiment 1 HG and the IV. Abteilung of the Fallschirm-Flak-Regiment HG.

The security duties were taken on by the newly created Begleit-Batallion Reichsmarschall, which was charged with additional special duties, i.e. the gathering of the replacements for training sections of the Fallschirmpanzerkorps HG and the German Volkssturm (militia).

In March 1945 the Begleit-Batallion Reichsmarschall was dissolved. The remaining units were wiped out in the fighting for Berlin.

	Panzer III	Panzer IV	Panzer V	Sturm-geschütz	Panzer-jäger
Sonderkraftfahrzeug	141/1	161/2	171 A	142	162/2
Gefechtsgewicht	23 t	25 t	45,5 t	21,8 t	25 t
Besatzung	5 Mann	5 Mann	5 Mann	4 Mann	5 Mann
Panzerung					
Front	50 mm	80 mm	80 mm	50 mm	80 mm
Seite und Heck	30 mm	30 mm	40 mm	30 mm	30 mm
Decke	10 mm	16 mm	16 mm	-	-
Motorleistung	300 PS	300 PS	600 PS	300 PS	300 PS
Geschwindigkeit (Straße)	40 km/h	40 km/h	46 km/h	40 km/h	40 km/h
Fahrbereich (Straße)	165 km	210 km	200 km	120 km	160 km
Länge (ohne Kanone)	5,48 m	5,93 m	6,88 m	5,40 m	6,15 m
Breite	2,95 m	2,88 m	3,42 m	2,95 m	2,92 m
Höhe	2,44 m	2,68 m	3,10 m	1,96 m	1,94 m
Kampfwagenkanone	7,5 cm+	7,5 cm	7,5 cm	7,5 cm	7,5 cm
Munitionsausstattung	87 Schuß	87 Schuß	79 Schuß	50 Schuß	76 Schuß

Die Panzer des Panzer-Regiments HG

+auch mit 5 cm-Kanone

The Panzers of the Panzer-Regiment HG
Panzer III, Panzer IV, Panzer V, assault guns, anti-tank guns.
Special vehicles; battle weight; crew; armor: front, sides and back, top; engine performance (PS+horse power); speed (on road); range (on road); length (without gun); width; height; guns (+ also with 5 cm gun); ammunition supply (Schuss'shell).

The Panzer III.

The Panzer IV was predominant in the Panzer-Regiment HG. To guard against limpet mines it could be equipped with 'aprons'. These were armor plates hung on the sides and mounted at the turret.

The Panzer IV.

The Panzer V "Panther" was an outstanding design and superior to all enemy tanks of its time. In the Fall of 1944, the I. Abteilung of the Fallschirm-Panzer-Regiment HG was equipped with it. The II. Abteilung was at the training grounds Grafenwoehr to be re-equipped with the "Panther" but did not return to the Korps before the capitulation.

The Sturmgeschutz III (assault gun).

Die Kanonen des Flak-Regiments HG				
	2 cm-Flak	3,7 cm-Flak	8,8 cm-Flak	10,5 cm-Flak (Eisb.)
Anfangsgeschwindigkeit der Granate (V_o)	835 m/s	820 m/s	830 m/s	900 m/s
Schußweite	4 800 m	6 500 m	14 860 m	18 000 m
Schußhöhe	3 700 m	4 800 m	11 000 m	13 000 m
Gewicht in Feuerstellung	450 kg	450 kg	5 000 kg	7 000 kg
Gewicht in Fahrstellung	770 kg	1 950 kg	7 200 kg	-

The guns of the Flak-Regiment HG
2 cm Flak; 3.7 cm Flak; 8.8 cm Flak; 10.5 cm Flak (railroad).
Initial velocity; shell range; shell elevation; weight in firing position; weight in mobile mode.

Of all the units, only the 14. (railroad) Flak-Batterie/RGG and the Fuhrer-Flak-Abteilung at the "Wolfsshanze" (wolf's den 'Fuhrer' headquarters), the latter being permanently built, were equipped with 10.5 cm gun.

In front a 3.7 cm gun, to the right the legendary "Achtacht" (8.8 cm gun).

The mounting of the 2 cm guns, as single or four-barrel, was varied: on special trailers, on trucks, or self-propelled on wheels or tracks.

The "Achtacht" had proven itself as a multipurpose gun. It was equipped with an armor plate to protect the crew during ground fighting.

The Defensive Battles in East Prussia

While in the middle of the process of organization, the Fallschirmpanzerkorps HG was moved, at the beginning of October 1944, by rail to East Prussia to stop, together with the divisions already in action there, the Soviet armies storming towards the Memel area and on to East Prussia from the north. For this purpose, the Korps was put into action north of the Russ river, the northern mouth of the Memel.

In the middle of October the Bolsheviks attacked along both sides of the road Wilkowischen-Ebenrode with strong tank units, their objective being Koenigsberg. In the sector of the 4. Armee (General Hossbach), units of the Fallschirmpanzerkorps HG had to take part in bloody fighting with Soviet tank spearheads which had pushed through to

towards the German civilian population. The Fallschirm-Panzer-Regiment HG and the Sturmgeschuetz-Abteilung HG, in particular, played a major role in preventing the planned breakthrough of the Soviet tanks into East Prussia and in pushing them back, once more, behind the Rominte river. One sergeant alone, with his "Panther", knocked out thirteen enemy tanks on October 20, 1944.

After the border battle in East Prussia died down, the Fallschirmpanzerkorps HG was stationed southeast of Guminnen from November 1944 to January 1945. In addition to its corps troops and its own two divisions, the Panzergrenadier-Division Grossdeutschland, the 21. and 61. Infantrie-Divisions and the 349. and 549 Volksgrenadier-Divisions were assigned to it. There was only static fighting. The Soviets had withdrawn their shock armies to reinforce them after the fighting which had, for them too, caused high losses, to reorganize them and to prepare them for the next wave of attacks. The 4. Armee, too, reorganized in anticipation of an attack by the Soviets. The Fallschirm-Panzer-Division 1 HG was withdrawn from the front as operative reserve, and on the order of the supreme commander of the Heer Mitte (central), separated from its own corps and moved, with the I./Fallschirm-Flak-Regiment HG and its supply units attached, to Poland, destination Radom.

When the Red Army began its anticipated major offensive along all of the Eastern Front on January 13, 1945, the units of the Fallschirmpanzerkorps HG remaining in East Prussia, together with the attached army divisions, entered into heavy, bloody battles high in casualties, which led to a staged retreat via Insterburg, Wehlau, Heilsberg, Landsberg, Kreuzberg, Zinten, Braunsberg to Heiligenbeil. Here, too, the Fallschirmpanzerkorps HG with its troops was encircled and a major part of it wiped out. The report of the armed forces stated on January 28, 1945 that, on the previous day, "in the sector of the Fall-schirmpanzerkorps HG 40 tanks were knocked out" and reported on February 10, 1945, "attempts by the Bolsheviks to break through in fierce tank attacks and repeated success in knocking out tanks in the sector of the Fallschirmpanzerkorps HG in the area of Landsberg-Kreuzberg".

In the early days of April 1945 the remaining units of the Fallschirmpanzerkorps HG were removed by way of the Frische Haff and the sea. Through Swinemeunde (sometimes also through Copenhagen) and Stettin, the remains were sent either to Berlin and Velten or directly to Silesia/Saxony where the Korps, from the middle of April, now again complete with its two divisions, was to be put into action, as the Fallschirm-Panzer-Division 1 HG had fought its way back to here within the 4. Panzerarmee (General Graeser) from Poland. The units of the Korps and of the Fallschirm-Panzergrenadier-Division 2 HG which had been brought from East Prussia temporarily to the Berlin area were now also moved to the new operational area.

As a visible sign of appreciation for courageousness in fighting with cold steel and hand-to-hand combat, the close combat clasp was created. It was awarded in three stages, in bronze, silver and gold after 15, 30 or 50 days in close combat. The picture shows the close combat clasp in silver, in the format acceptable today, without swastika.

```
                Aus einem Antrag des I./Fsch.Pz.Gren.Rgt. 2 HG
                 auf Verleihung der 2. Stufe der Nahkampfspange

         Nahkampftage laut Regimentsbefehl Fsch.Pz.Gren.Rgt. 2 HG

          1.    20. 6.44    Kämpfe um Chiusi
          2.    21. 6.44    Abwehrkämpfe um Chiusi
          3.    24. 6.44    Abwehrkämpfe bei S. Savino
          4.    28. 6.44    Kämpfe um Statione d. Montepulciano
          5.    29. 6.44    Abwehrkämpfe bei Abbadia
          6.    10. 7.44    Kämpfe bei Ambra
          7.     6. 8.44    Angriff auf Emilow-Michalowek
          8.     8. 8.44    Angriff auf Michalow
          9.     9. 8.44    Einnahme von Michalow
         10.    11. 8.44    Panzervorstoß auf Studzianki
         11.    12. 8.44    Einnahme von Studzianki
         12.    13. 8.44    Abwehrkämpfe um Studzianki
         13.    20. 8.44    Abwehrkämpfe um Michalow-Rogozek
         14.    23. 8.44    Abwehrkämpfe an der Radomka
         15.    20. 9.44    Zerschlagung eines russischen Brücken-
                            kopfes bei Warschau
         16.     2.10.44    Abwehr feindlicher Angriffe nördlich
                            Warschau
         17.    12.10.44    Abwehrkämpfe bei Usspelken
         18.    13.10.44    Abwehrkämpfe bei Usspelken
         19.    16.10.44    Angriff auf Werzenhof bei Usspelken
         20.    20.10.44    Angriffskämpfe um Bissnen
         21.    21.10.44    Kämpfe bei Rodebach
         22.    22.10.44    Kämpfe bei Rodebach
         23.    23.10.44    Kämpfe ostwärts Trakehnen
         24.    24.10.44    Kämpfe um Trakehnen
         25.    25.10.44    Kämpfe um Trakehnen und Goltzfelde
         26.    26.10.44    Kämpfe um Guddin
         27.    30.10.44    Kämpfe um Guddin
         28.    25. 1.45    Gegenangriff auf Jägersdorf
         29.    26. 1.45    Gegenangriff auf Gut Dommerau
         30.    27. 1.45    Abwehrkämpfe um Gut Hansfelde
```

'From a request of the I./Fsch.Pz.Gren.Rgt. 2 HG to award the second stage of the close combat clasp. Days of close combat according to regimental order Fsch.Pz.Gren.Rgt. 2 HG: 1. Battle for Chiusi, 2. Defensive Fighting for Chiusi, 3. Defensive fighting near S. Savione, 4. Battle for Statione d. Montepulciano, 5. Defensive fighting near Abbadia, 6. Battle near Ambra, 7. Attack on Emilow-Michalowek, 8. Attack on Michalow, 9. Capture of Michalow, 10. Panzer assault on Studzianki, 11. Capture of Studzianki, 12. Defensive fighting for Studzianki, 13. Defensive fighting for Michalov-Rogozek, 14. Defensive fighting at the Radomka River, 15. Destruction of a Russian bridge-head near Warsaw, 16. Repulse of enemy attacks north of Warsaw, 17. Defensive fighting near Usspelken, 19. Attack on Werzenhof near Usspelken, 20. Offensive fighting for Bissnen, 21. Fighting near Rodebach, 22. Fighting near Rodebach, 23. Fighting east of Trakehnen, 24. Battle for Trakhnen, 25. Battle for Trakehnen and Goltzfelde, 26. Battle for Guddin, 27. Battle for Guddin, 28. Counter attack on Jagersdorf, 29. Counter attack on Gut Dommerau, 30. Defensive fighting for Gut Hansfelde'.

The Pak (anti-tank guns), in particular the excellent 7.5 cm Pak 40, had a major part in the high knock-out success.

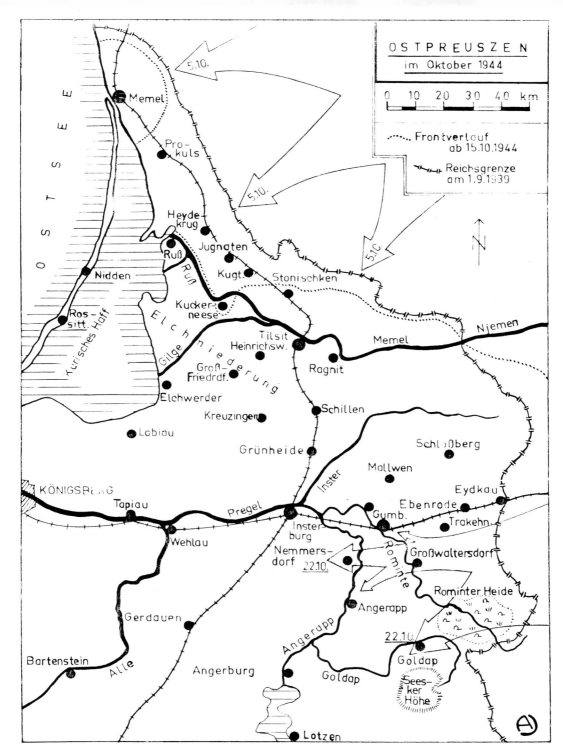

East Prussia in October 1944.

Front line, border of the Reich on 9/1/39.

With two strong spearheads, one from the northeast towards the mouth of the Memel river, the other from the east directly towards Koenigsberg, the Soviets started the conquest of East Prussia. Thus it came to the border battles in East Prussia.

The I./Fallschirm-Panzer-Regiment HG, just equipped with the excellent new "Panther", was one of the strongest fighting units which resisted the Soviet tank packs in the border battles in East Prussia and knocked out 99 tanks and 43 anti-tank guns as well as numerous other weapons in four days. The 1. Kompanie knocked out 47 enemy tanks and 30 anti-tank guns in six days; Feldwebel Bowitz of the 4. Kompanie, alone, within a few hours, knocked out a further 13 enemy tanks.

This T34, too, succumbed in Nemmersdorf.

In November 1944, the supreme commander of the Luftwaffe visited his Korps in East Prussia. At Gut Austinshof, where the Korps staff had its quarter, he was informed of the situation in the sector of the Korps. From the left: Reichsmarschall Göring, the chief of the general staff Oberstleutnant i.G. von Baer, the orderly officer Leutnant Kleine-Sextro, the commanding general Generalmajor Schmalz and his quarter master Major i.G. Gruen.

On November 3, 1944, the Panzerkampf (tank battle) badge was introduced. With silver oak garland it was awarded to members of the Panzer and tank reconnaissance units who had excelled in three battles on three different days, and with black oak garland, to be awarded to members of the Panzergrenadier units who had taken part in three assaults in the front line with weapons in hand on three different days. The picture shows the badge in today's approved format, without swastika.

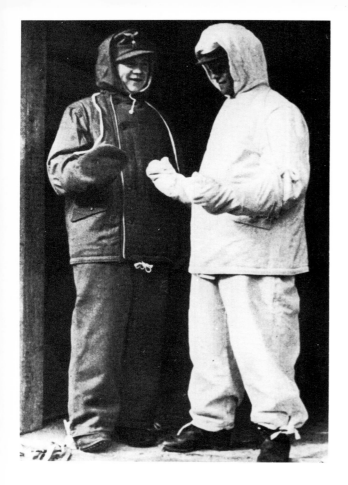

In time, before the start of the hard winter 1944/45, the front troops were equipped with the new winter fighting suits which, depending on weather conditions, could be worn with the gray-green or white side out.

At the end of December, when severe frost set in and heavy snow fell soon thereafter, the soldiers sought, during the static fighting, shelter from the cold in deep bunkers. Still, alert sentries watched the approaches to detect approaching enemy scouts in time.

To save the dwindling fuel supply, missions, as in this photo, were carried out whenever possible with horse drawn carts. The Fallschirmpanzerkorps has become the Fallschirmpanje (horse) korps.

The Sturm-(assault) battalion of the Fall-schirmpanzerkorps HG had an authorized strength of 974 men. From September to December 1944 the Battalion lost 606 men, dead and wounded; that is almost two-thirds of authorized strength. These hard figures show how great the losses were. Bigger losses, still, occurred in the following months, in particular in the encircled Heiligenbeil.

Wherever the enemy had overrun our positions, the "Panthers" went on the counteroffensive, often with Grenadiers mounted.

It was a great day for the Korps: more Knight's Crosses. At Gut Austinshof soldiers receive their awarded decoration in mid-December 1944: seven Knight's Crosses and one German Cross in Gold. They are (from left): Hauptmann Stuchlik, Oberleutnant Kraus, Leutnant Wallhaeusser, Unteroffizier Grunhold, Gefreiter Steets, Gefreiter Plapper and Oberfeldebel Bowitz. Hauptmann Renz, fallen in the meantime, should also have been there.

Generalmajor Schmalz in November 1944 with several holders of the Knight's Cross of the Korps. From the left, first row: Oberst von Necker, Generalmajor Schmalz, Oberst Soeth; second row: Major i.G. Gruen, Oberstleutnant i.G. von Baer, Major Rossmann; third row: Oberfeldwebel Kulp, Major Rebholz, Major Ostermeier, Oberfeldwebel Schlund, Oberstleutnant Kluge.

The Fallschirmpanzerkorps HG in the border battle in East Prussia. Top arrow: Soviet tank spearheads from 10/19-10/22, 1944. Second arrow: German units. HG Fallschirmpanzerkorps HG. 5.Pd 5th Panzerdivision. FBG Fuhrerbegleitbrifade (Fuhrer escort brigade). AK Armeekorps. Dotted line Front line Nov. 44 to Jan. 45.

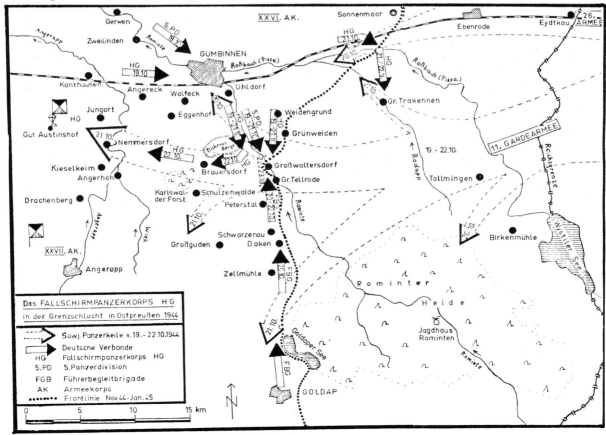

Das FALLSCHIRMPANZERKORPS HG
in der Grenzschlacht in Ostpreußen 1944

- - - →	Sowj. Panzerkeile v. 19.- 22.10.1944
➤	Deutsche Verbände
HG	Fallschirmpanzerkorps HG
5.PD	5.Panzerdivision
FGB	Führerbegleitbrigade
AK	Armeekorps
···········	Frontlinie Nov.44-Jan.45

0   5   10   15 km

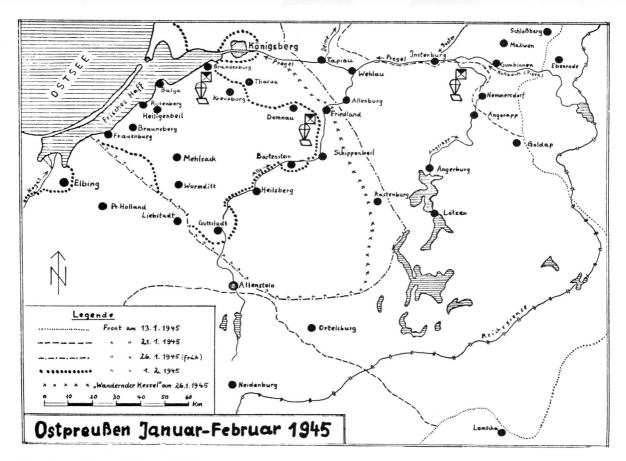

'East Prussia January-February 1945'.

The encirclement at Heiligenbeil, March 1945.

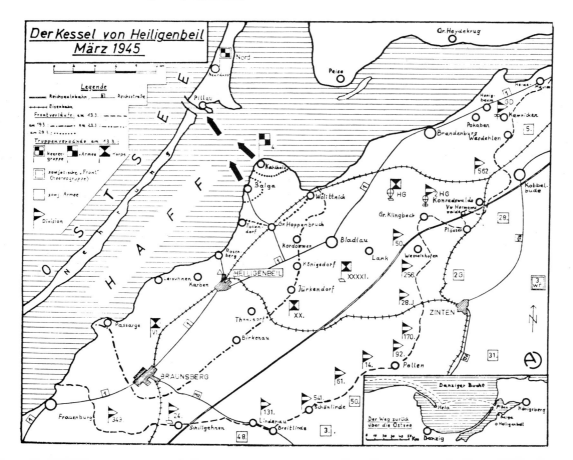

Oberfeldwebel Kulp fell on February 5, 1945 near Sollnicken/Zinten as platoon leader in the 13./Fallschirm-Panzergrenadier-Regiment 4 HG.

'Temporary ownership certificate. The Führer and commander-in-chief of the armed forces has awarded the Feldwebel Karl Kulp the Knight's Cross of the Iron Cross. Headquarters of the supreme commander of the Luftwaffe, September 12, 1944. The chief of personnel and national-socialist leadership of the Luftwaffe'.

The units of the Korps, cut off in the encirclement of Heiligenbeil, were lacking many things, such as document forms for the awarding of medals. Thus, improvised substitute documents had to be issued. The lack of officers is noticeable here, too. The senior noncoms lead companies.

'To the Unteroffizier Gustav Kretschmer, Gablonz 11a, Gurtlerstyr.24. In the name of the Supreme Commander of the Luftwaffe you are awarded, effective 2/17/45, the Panzerkampf (tank battle) abzeichen. To save paper and for reasons of the major action, no certificate has been issued for this award. I offer most sincere congratulations on the award. Valid in place of a certificate'.

'Marching Order (in lieu of the armed forces travel document). The (rank, name, first name) of the above mentioned unit is in transit from Pillau to Berlin-Velten and has orders to report in the fastest possible way to the Fsch.Pz. Gren.Div.2 Hermann Göring IIb. Reason: move of the Fsch.Pz.Gren.Div. 2 HG to the center of the Reich. The Marching Order is valid, in lieu of the armed forces travel document as a ticket on the German railroad.'

Where an organized transport of units was no longer possible, soldiers were issued makeshift marching orders to travel on their own.

The Replacement and Training Brigade HG

With the expansion of the Fallschirm-Panzer-Division HG to the Korps, the Replacement and Training Regiment HG in Holland was expanded to the Fallschirm-Panzer-Ersatz (replacement) and Ausbildungs-(training) Brigade HG. With personnel support of the old regiment in Holland, the Fallschirm-Ersatz and Ausbildungs Regiments 1 and 2 HG were formed in Rippin, West Prussia. Parts of the regiment remained in Holland, however. It had received much personnel from disbanded flying units and ground units of the Luftwaffe for re-training.

In mid-January 1945, when the rapidly approaching front reached the West Prussian bases of the Ausbildungs-Brigade, it retreated, being hardly mobile by itself, fighting together with the 83. Infantrie-Division, by way of the Drewenz position and Briessen as well as Rehden to Graudenz and was there put into action as a fortress garrison.

The Brigade, with the exception of the training and administrative personnel, consisted mainly of men drafted only a few weeks previously, not yet completely trained, whose deployment to the front at that stage of training would otherwise never have been considered. Through the personal influence of the brigade commander, at least those drafted last could be sent back to the Reich.

The few survivors of the Brigade were taken prisoner of war by the Soviets when the fortress Graudenz capitulated on March 7, 1945. The Fallschirm-Panzer-Ersatz and Ausbildungs-Brigade HG and thus ceased to exist.

The route of the Fallschirm-Panzer-Ersatz and Ausbildungs-Brigade HG from Rippin to Graudenz. Before the encirclement of the fortress Graudenz, the Regiment 1, Hauptmann Findeis, was removed from the Brigade and put into action in the fighting down river of the Weichsel where it was lost.

The commander of the Ersatz and Ausbildungs-Brigade HG, Oberst Meyer, who was taken prisoner of war by the Soviets with his remaining men, during an attempt to break out, on March 8, 1945.

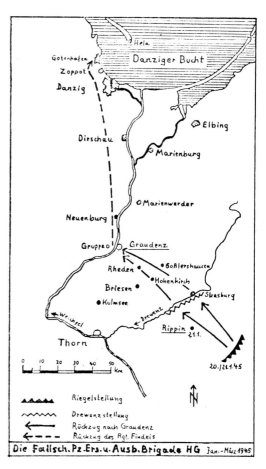

In the last war year, those born in 1927 and 1928, and on March 5 even those born in 1929, were drafted. They were still half children as they had to join the war.

Recruits moving out after swearing in ceremony.

The Fallschirm-Panzer-Division 1 HG in Poland, Silesia and Saxony

The Fallschirm-Panzer-Division 1 HG, coming from East Prussia, could not reach its destination of Radom since it was already in enemy hands, and had to be re-directed to the Litzmannstadt (Lodz) area. There it was unloaded and attached to the Panzerkorps Grossdeutschland which had also been brought in from East Prussia. Both corps has incomprehensibly been torn apart. The Panzergranadier-Division Grossdeutschland which had remained in East Prussia was there attached to the Fallschirmpanzerkorps HG.

The Panzerkorps Grossdeutschland, with its divisions, cut off rom all supply routes and vehemently beset by the enemy, fought its way westward, roughly along the line Kalisch-Ostrowo-Krotoschin-Silesian border near Guhrau as a 'moving pocket' to the Oder river. It was crossed by the Fallschirm-Panzer-Division 1 HG during the night of January 31/February 1, 1945 on a hastily built bridge near Oberbeltsch; located between Glogau and Koeben. The further retreat led via Sprottau-Sagan to the Lausnitzer Neisse near Muskau where static fighting then occurred.

In March 1945 the Fallschirm-Panzer-Division 1 HG initially fought in the Lauban area, then, in the middle of that month, in the battle of Upper Silesia in the Neisse-Grottkau area, and finally, in mid-April, north of Goerlitz. Here, near Koderdorf, the Panther unit of the Fallschirm-Panzer-Regiment HG knocked out 43 Polish tanks within 20 minutes and, additionally, captured twelve more undamaged which, furnished with the cross insignia, were immediately incorporated into the Panzer-Regiment. At the end of April the Division took part in the fighting for Bautzen which led to the re-liberation of this city.

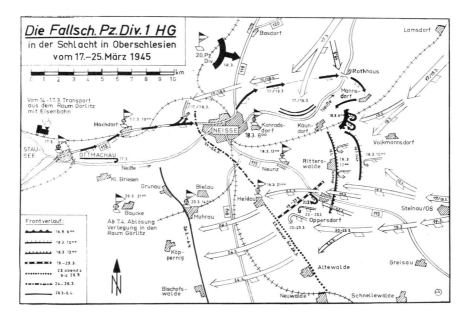

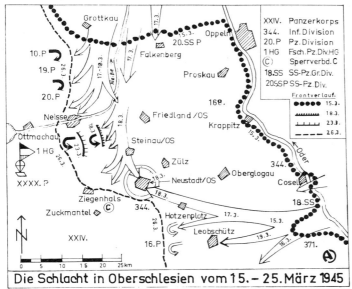

'The Fallsch.Pz.Div. 1 HG in the battle in Upper Silesia from 17-25 March 1945'.

'The battle in Upper Silesia from 15-25 March 1945'.

The Panzer-Regiment was the backbone of the Division, in particular the "Panthers" of the I. Abteilung. The "Panther" had an extremely effective 7.5 cm gun, caliber length 70, with enormous penetration power.

Among the best commanders of the Panzergrenadiers were Major Brigel (left), leader of the Regiment 2, and Major Moergel, leader of the Regiment 1, who fell, in the final days of the war, near Dresden. Both had been awarded the Knight's Cross.

In Upper Silesia, too, as so often, the Panzergrenadiers carried the main burden of the fighting.

The commander of the Panzer-Regiment, Oberstleutnant Rossmann, with two of his officers during the battle in Upper Silesia.

A short break in the march is used to make a sandwich and to have a look at the front newspaper.

Return by rail from Upper Silesia back to the Goerlitz area.

'Temporary certificate of ownership. The Fuhrer and commander-in-chief of the Wehrmacht has awarded the Major Karl Rossmann the Oak Leaves to the Knight's Cross of the Iron Cross on February 1, 1945. Berlin, February 14, 1945. The chief of the Luftwaffe personnel administration. General of the pilots'.

The Last Battles of the Fallschirmpanzerkorps HG in Saxony

In mid-April the general command and the corps troops of the Fallschirmpanzerkorps HG, together with the Fallschirm-Panzergrenadier-Division 2 HG, after a hasty reinforcement and only marginally equipped, arrived in the new operational area of Hoyerswerda-Bautzen. Soon after, the Fallschirm-Panzer-Division 1 HG was also re-attached to its own Korps.

Fighting as a unit, the Fallschirmpanzerkorps HG achieved another (the last) success near Koenigsbrueck: there, it repelled the Soviet tank spearhead, including the 1 Polish Division, with great losses.

During the subsequent fighting between Dresden and Grossenhain, in the evening of May 8, 1945, the order arrived that on May 9 at 0 hours 1 minute the armistice would begin. The units were ordered, as far as the vehicle and fuel situation allowed, to immediately evade to the south through Czechoslovakia and to the west in order to reach American custody. However, the fast advance of the Soviet tank spearheads of Marshal Konjew from Berlin in the direction of Prague closed the encirclement which still contained the majority of the Fallschirmpanzerkorps HG. Only few units succeeded in fighting their way, partly also through the Czech partisan forces, to American captivity. Most had to begin the march into Soviet imprisonment.

Of the many prisoners who had to perform compulsory hard labor on insufficient nourishment in primitive camps in the Soviet Union, a majority could not endure the hardships and did not see the homeland again. Many were sentenced to long-term hard labor (mostly 25 years) because, their only 'crime', they had belonged to an elite unit.

Those who were allowed to return home after many years imprisonment (the majority at the end of 1949) were often marked by illness and found a fatherland in ruins and occupied by foreign troops. Those from the German eastern territories could not return there and had even greater difficulties to again build an existence, this was, in most cases, only possible after years of determined endeavors and many privations.

**Einsatzdaten des Stabes
Fallschirm-Panzergrenadier-Regiment 4 HG
im Jahre 1945**

13.1.45	Beginn des Großangriffes der Sowjets bei Schloßberg-Ebenrode, von 7 bis 11 Uhr Trommelfeuer
19.1.45	Verlegung von Wolfseck über Nemmersdorf nach Königsgarten, 10 km westlich von Nemmersdorf
21.1.45	Verlegung nach Linnemarken bei Königsgarten
22.1.45	Verlegung zum Fliegerhorst Uhlenhorst bei Nordenburg
23.1.45	Verlegung nach Altendorf
24.1.45	Verlegung nach Heinrichsdorf (Flakstand)
25.1.45	Wittenberg, südlich Königsberg, Fliegerhorst Jesau
27.1.45	Mahnsfeld, Bombenangriff
28.1.45	Willmsdorf
29.1.45	Vorwerk bei Zinten
1.2.45	Gut Korschellen
8.2.45	Nemritten
9.2.45	Dösen
10.2.45	Wesselshofen, südlich der Autobahn
12.2.45	Lank
22.2.45	Verlegung in den Wald bei Baumgarten/Lank
23.2.45	Verlegung nach Bladiau
25.2.45	Verlegung nach Ludwigsort
8.3.45	Major Druckenbroth beerdigt in Ludwigsort
15.3.45	Verlegung nach Bolbitten, nördlich von Bladiau, fast am Frischen Haff
16.3.45	Verlegung nach Rensegut, 6 km südlich von Balga
18.3.45	Oblt. Schink gefallen
19.3.45	Verlegung nach Lindenberg, Gut Unruh
27.3.45	Major Stauch gefallen, Ostpreußen wird aufgegeben
30.3.45	Abmarsch zur Sammelstelle HG in Klein Blumenau, an der Straße Pillau-Königsberg
31.3.45	In Baracke Elenskrug
6.4.45	Ab von Elenskrug nach Pillau, Kurfürstenkaserne
7.4.45	Warten auf Verladung im Hafen Pillau, Fliegerangriff
8.4.45	16 Uhr Verladen auf Schiff "Orestes", 2000 Soldaten und 2500 Flüchtlinge. Um 22.15 Uhr Auslaufen
9.4.45	Geleitzug 3 Transportschiffe, 2 Sicherungsboote, 2 Minensuchboote, vor Hela gelegen wegen U-Boot- und Schnellbootgefahr. Russischer Bombenangriff auf Geleitzug
12.4.45	Ankunft in Kopenhagen. Um 12.30 Uhr Ausladen
13.4.45	16 Uhr wieder Einladen auf "Orestes"; 18.30 Uhr Auslaufen
14.4.45	Ankunft in Swinemünde, um 8 Uhr Ausladen. Um 12 Uhr weiter mit Flugsicherungsboot nach Stralsund. Um 20 Uhr Ausladen, Quartier in Franken-Kaserne
15.4.45	Abt. IIa/b um 4.25 Uhr mit Eisenbahn über Angermünde nach Eberswalde, Nachtquartier
16.4.45	Mit Eisenbahn um 5.30 Uhr Weiterfahrt nach Berlin, Ankunft 5 Uhr, Weiterfahrt um 12 Uhr in Richtung Görlitz. In Cottbus Fliegerangriff, danach weiter nach Senftenberg
17.4.45	Um 5.10 Uhr Weiterfahrt nach Wiednitz zur Division und danach weiter nach Großweidau bei Hoyerswerda
19.4.45	Verlegung nach Bernsdorf, dann nach Cosel bei Schwepnitz
20.4.45	Verlegung über Königsbrück nach Rittergut Rennersdorf bei Stolpen
21.4.45	Verlegung nach Polenz bei Neustadt/Sachsen
26.4.45	Verlegung nach Nieder-Putgkau bei Bischofswerda
28.4.45	Verlegung nach Bretnig
1.5.45	Hitlers Selbstmord wird bekanntgemacht als "gefallen"
2.5.45	Berlin kapituliert
5.5.45	Verlegung nach Polenz
7.5.45	Verlegung nach Oelsen bei Liebstadt-Glashütte/Erzgebirge
8.5.45	Befehl zur Kapitulation, die am 9.5.45 um 0,01 Uhr in Kraft tritt

'Dates of actions of the staff Fallschirm-Panzer-grenadier-Regiment 4 HG in the year 1945.
1/13/45 Start of the major Soviet offensive near Schlossberg-Ebenrode, from 7 to 11 a.m. barrages
1/19/45 Shift from Wolfseck via Nemmersdorf to Koenigsgarten, 10 km west of Nemmersdorf
1/21/45 Shift to Linnemarken near Koenigsgarten
1/22/45 Shift to air base Uhlenhorst near Nordenburg
1/23/45 Shift to Altendorf
1/24/45 Shift to Heinrichsdorf (Flak position)
1/25/45 Wittenburg, south of Koenigsberg, air Jesau
1/27/45 Mahnsfeld, bomb attack
1/28/45 Willmsdorf
1/29/45 Vorwerk near Zinten
2/1/45 Gut Korschellen
2/8/45 Nemritten
2/9/45 Doessen
2/10/45 Wesselhof, south of the Autobahn (major highway)
2/12/45 Lank
2/22/45 Shift to the woods near Baumgarten/Lank
2/23/45 Shift to Bladiau
2/25/45 Shift to Ludwigsort
3/8/45 Major Druckenbroth buried in Ludwigsort
3/15/45 Shift to Bolbitten, north of Bladiau, almost at the Frische Haff
3/16/45 Shift to Rensegut, 6 km south of Balga
3/18/45 Oblt. Schink fallen
3/19/45 Shift to Lindenberg, Gut Unruh
3/27/45 Major Stauch fallen, East Prussia abandoned
3/30/45 March to the assembly area HG in Klein Blumenau, on the road Pillau-Koenigsberg
3/31/45 In the cabin at Elenskrug
4/6/45 From Elenskrug to Pillau, Kurfuersten barracks
4/7/45 Waiting for loading in Pillau harbor, air raid
4/8/45 1600 loading again onto vessel "Orestes", 2000 soldiers, 2500 refugees. Departure at 2215 hours
4/9/45 Convoy 3 transport vessels, 2 security vessels, 2 mine sweepers, waiting at Hela because of danger of submarines and speed boats. Russian bomb attack on convoy
4/12/45 Arrival at Copenhagen. Unloading at 1230
4/13/45 1600 loading again onto "Orestes"; 1830 departure
4/14/45 Arrival at Swinemuende, unloading at 0800. At 1200 onwards by air rescue boat to Stralsund. Unloading at 2000, quarters at Franken barracks
4/15/45 Abt.IIa/b at 0425 by train via Angermuende to Eberswalde, night quarters
4/16/45 Continuation by train at 0530 to Berlin. Arrival at 0800, on towards Goerlitz at 1200. In Cottbus air raid, then on to Senftenberg
4/17/45 At 0510 continuation to Wiednitz to the Division, and then on to Gorssenweidau near Hoyerswerda
4/19/45 Shift to Bernsdorf, then to Cosel near Schwepnitz
4/20/45 Shift via Koenigsbrueck to Rittergut Rennersdorf near Stolpen
4/21/45 Shift to Polenz near Neustadt/Saxony
4/26/45 Shift to Nieder-Putgkau near Bischofswerder
4/28/45 Shift to Bretnig
5/1/45 Hitler's suicide is announced as 'fallen'
5/2/45 Berlin capitulates
5/5/45 Shift to Polenz
5/7/45 Shift to Oelsen near Liebstadt-Glashuette/Erz Mountains
5/8/45 Order to capitulate effective 5/9/45 at 0001

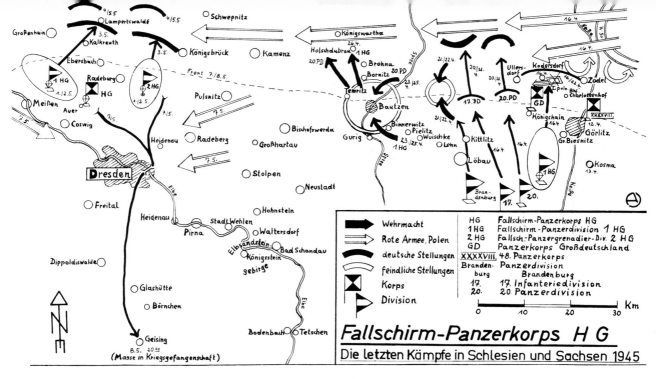

'Fallschirm-Panzerkorps HG. The last battles in Silesia and Saxony 1945.

The remains of the Fallschirmpanzerkorps HG were extricated from the encirclement of Heiligenbeil by sea and were to be re-enforced and reassembled as a complete Korps with it own Panzer-Division. Instead of in the Herzberg/Riesengebirge area, the re-enforcement occurred north of Berlin. But not all units could be refitted there, some were directly moved to the new operations area.

One of the last documents on the Fallschirmpanzerkorps HG. The expected decision was obviously not made, days later, on April 23, 1945, Göring was removed from all his offices. The general command of the Korps had not received the order to reorganize, it had other problems, in any case, than to reorganize its units which, bled dry and at the end of their strength, doggedly resisted the continued onslaught of the Soviets. It became apparent that the end was near.

Re: Re-enforcement of Fallschirm-Panzerkorps "Hermann Göring". Operations unit command group has submitted to the Fuhrer the following regarding the re-enforcement of the Fallschirm-Panzerkorps HG: 1.) Adding personnel: After arrival of the general command of the Pz. Korps "Hermann Göring" and the Fallsch.Pz.Div. "Hermann Göring" 2 planned for short term in the area north of Bernau. 2.) Materiel replenishment: Suggested in the area west of Hirschberg. Are is particularly favorable since supply routes from the factories would be greatly shortened (handguns from Braenn, guns from Skoda). Further advantages: Tactical assembling of all of the Fsch.Pz.Korps "Hermann Göring 1" is in action in the same area. 3.) Time frame of the resupply: Addition of personnel is planned to begin on 4/12, to be complete 4/19. Vehicles have to be provided by the Luftwaffe since production of new vehicles has come to an almost complete standstill. To expedite the transport of weapons and equipment, assistance from the transport units of the Luftwaffe is greatly required. It is noted: (note from the Fuhrer situation chief Luftwaffe) 1.) Fuhrer agreed with the proposal. 2.) The OKH (supreme command army) has until now been responsible for the resupply of the Pz.Korps "Hermann Göring" with vehicles.

'Note to file No. 20. Re: Re-enforcement Fsch.Pz.Korps HG 1.) With the issuance of the new organization Pz.Div.45, it is equally ordered to organize all Panzer and Pz.Gren.Div. in the same manner by 5/1/45. Thus, the Pz.Div. equals the Pz.Grem. Division. 2.) To refit the Fsch.Pz.Korps HG there are two possibilities: a) Provisions with Panzers not possible: establishment of 2 Fsch.Jg.Div.45, or 1 Fsch.Pz.Div. and 1 Fsch.Jg.Div.45. Organization of Korps units remain unchanged. 3.) It is requested that the above possibilities for the refit of the Fsch.Pz.Korps HG be submitted to the Reichsmarschall for decision.'

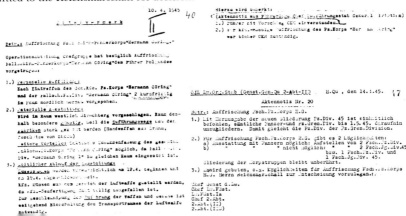

Fighters and fighter bombers caused the troops many problems, and many a Panzer, vehicle and gun became their victim.

Without pause, enemy aircraft attacked not only the troops in the battle areas, but also knocked out the supply routes, in particular the traffic arteries.

Towards the end of the war, companies (in emergencies even battalions) were led by sergeants and staff sergeants. Tested and proven in many battles, the sergeants (with their long experience on the front) led the men entrusted to them, and offered them time and again a personal example of courage fearlessness.

The Pak was victorious! The front wall of the T34 is cleanly pierced.

A 7.5 cm Pak in battle with an approaching Soviet tank.

The End of the Other HG Units

The replacement and training Regiment HG, which had remained in Holland, reached, despite the losses to the newly created training Brigade HG in Rippin, a strength of over 12,000 men because former aircraft crew and members of disbanded units of the Luftwaffe ground support organization were attached to it. In existence were a Grenadier-Battalion, a Pionier-Batallion and an Abteilung each of Artillerie with Flak and Panzers with tank destroyers, each unit at about regiment strength.

The training Regiment was soon put into action at the front and suffered, because of a lack of training in ground fighting and of experience, considerable losses. Based on the high numbers of personnel, the Sturm-Regiment z.b.V. HG was formed in September 1944 and the Fallschirm-Jager-Regiment 31 in February 1945. The former fought its way through Holland and Belgium back to the Rhineland. Its remains were taken prisoner in the Bonn-Bad Godesberg area.

After the loss of the Fallschirm-Panzer-Ersatz (replacement) and the Ausbildungs (training) Brigade HG in Graudenz, a new Brigade with the number 2 and the Regiments 3 and 4 was formed near Joachimstal, and soon after moved to the Oder front to see action in the area of Eberswalde-Angermuende, from where it fought its way back to Mecklenburg under heavy losses.

The soldiers of the central staff HG, posted in Berlin-Reinickendorf, were deployed in the defense of Berlin and lost. The barracks area sustained considerable damage.

In March 1945, parts of the Fallschirm-Panzer-Regiment HG (mostly of the II. Abteilung) were at the training grounds at Grafenwoehr for replacement and re-equipment with "Panthers". These units were brought into action within the hastily created Panzergruppe Grafenwoehr in the area south of Nuremberg and perished in the vicinity of Neumarkt, Allersberg, Hipoltstein, Alfershausen and Greding.

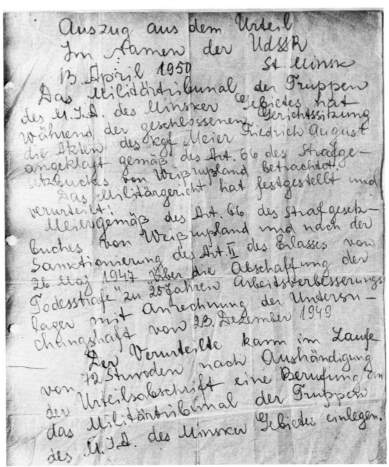

A special document. It is a thin sheet of onion skin paper, written with pencil. it speaks for itself.

Extract from the court decision. In the name of the USSR. April 13, 1950. St. Minsk. The military tribunal of the troops of the Minsk area had examined, in closed session, the files of the prisoner of war Meier, Friedrich August, charged in accordance with article 66 of the criminal code of White Russia. The military tribunal had deliberated and sentences: Meier, in accordance with article 66 of the criminal code of White Russia, and after the sanctioning of article II of the Directive of May 26, 1947 on the abolition of the death penalty, to 25 years labor camp with the detention from December 23, 1949 being deducted from the sentence. The convict may, within 72 hours of the handing down of the copy of the court findings, commence an appeal with the military tribunal of the Minsk area'.

From the armed forces report of May 9, 1945: "...Since midnight the weapons are silent now on all fronts. By order of the Grossadmiral (admiral of the fleet), the armed forces have ceased the fight which had become hopeless. thus, the almost six year honorable struggle is at an end. It has brought us great victories, but also grave defeats. The German armed forces succumbed, in the end, honorably to an enormous superiority..."

Our barracks at Berlin-Reinickendorf sustained heavy damage from Allied air raids. Quarters, storage buildings, garages, gun and ordnance parks, and repair shops were in ruins, the horses of the cavalry platoon, killed by bombs, lie in front of their stable.

The remains of the "Schweine-Kantine" (canteen).

The Barracks in Berlin-Reinickendorf Today

Our barracks in Berlin-Reinisckendorf is now located in the French sector and is named, in memory of Napoleon's entry into Berlin after his victory at Jena and Auerstedt on October 27, 1806, QUARTIER NAPOLEON. The barracks are the seat of the French commander and houses the French occupation forces in the strength of a brigade.

If one visits the barracks today, one finds them rather strange compared to the past. Some buildings which were completely destroyed by bombs have not been replaced. The area either remained empty or a new building, often in a different style, has been built there, such as, i.e. the domestic offices of the former III. (Scheinwerfer) Abteilung which housed the "Schweine-Kantine". At that spot, the office building for the French commander has been placed. In the area of the former Wach-Batallion, additional new buildings in accordance with the French requirements have been erected. A palace of culture and a garrison church have also been built within the barracks compound. Almost all official buildings have received window grates, and some grass areas were sacrificed for park areas. The numerous, luxuriously grown bushes, shrubs and trees, which were planted later, have, after more than forty years, given a different image to the whole landscape.

The Tegel parade ground, our former "playground" where many a drop of sweat was shed, is today the airport Tegel and is blocked off by a fence. The barracks now has a rail connection from the airport to the freight station at Reinickendorf.

Where the wooded area between the west border of the barracks and the Hohenzollern canal,there now is a housing estate for French families, the streets have French names.

Entry by strangers to the barracks is, for reasons of security, basically not possible. Only on the French national holiday, July 14, is the populace of Berlin invited to the 'Day of the Open Door' and has then the opportunity to enter the barracks and witness the French troops on parade.

The gateway to the Quartier Napoleon in 1960. The lattice in the center still carries the coat of arms of the house of Göring, but, newly placed above it: Napoleon on horseback. The green victory garland is bound (as is the Fallschirmschuetzenabzeichen (parachute gunner badge)) half each of laurel and oak leaves.

'Our barracks in Berlin-Reinickendorf as Quartier Napoleon'. Legend: barracks compound, living quarters for French soldiers' families, border to airport Berlin-Tegel, swimming pool, green areas, schools for French children.

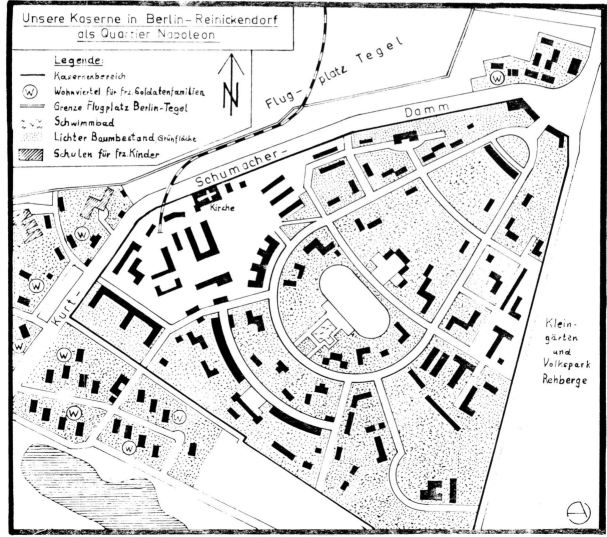

Unsere Kaserne in Berlin-Reinickendorf als Quartier Napoleon

Legende:
— Kasernenbereich
(W) Wohnviertel für frz. Soldatenfamilien
= Grenze Flugplatz Berlin-Tegel
Schwimmbad
Lichter Baumbestand, Grünfläche
Schulen für frz. Kinder

Flug-platz Tegel

Damm

Schumacher-

Kirche

Kurt-

Klein-gärten und Volkspark Rehberge

A second birth certificate. Returned home after 10 years of war duty and imprisonment. For many a new, harsh and laborious life begins. The will, to re-build the destroyed fatherland, provides the strength to overcome all obstacles and difficulties.

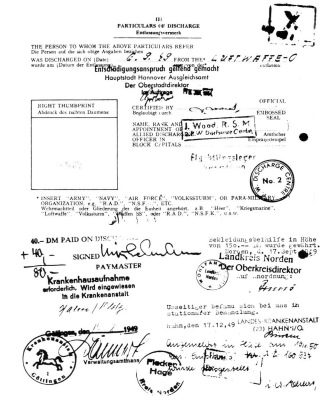

In the honor grove of the Panzer school of the German Federal Armed Forces in Munster, each of the Panzer and Panzer-Grenadier-Divisions of the former armed forces, has a boulder which, as its memorial stone, carries the divisional insignia on a bronze plate. Every year, on Remembrance Day, the Kampftruppenschule 2 (combat school) organizes a ceremony which includes the laying of a wreath. Representatives of the association of comrades of the Fallschirmpanzerkorps, also, lay wreaths at our two divisional boulders.

Insignia on the occasion of the founding of the association of comrades.

Fostering the Tradition

The terrible plight of the years following the war in the homeland which had shrunk so much in 1945, the anxiety for the daily bread, the defamation of the German soldier by the occupation forces, by the authorities, in press and radio, the pressure of a total occupation of the country, all these were reasons which made a meeting of old soldiers initially impossible. Even when the Western Allies demanded a German defense contribution in the early 1950's, German public opinion was still predominately hostile to soldiers and the military.

It required a good deal of courage when in 1952, the former Unteroffizier Udo Fischer called for a meeting of veterans, in particular those of the former HG units. He organized a meeting in Wassertruedingen/Mittelfranken, which brought together some 500 comrades from all of the then-occupied zones, even one from Berlin on a bicycle. Also attending, and joyfully welcomed, was the former commanding general of our korps, Generalleutnant (ret.) Schmalz who had only returned from imprisonment two years earlier.

The main concern of those attending was to establish the fate of comrades missing in action. Even at this first meeting 16 cases could be clarified. A firm association, however, was not yet formed in Wassertruedingen.

Only when two years later another former Unteroffizier, Adolf Gefeke, together with comrades from Celle and Alfred Otte of Hanover, organized another meeting and some 300 former HG members from the Federal Republic got together, the Kameraschaftsbund Fallschirmpanzerkorps (association of comrades) was formed on June 27, 1954, with the main objective: determining the fate of those missing in action. This objective has been successfully pursued by the Kameradschaftsbund, it continues to be pursued, there are still more than 8,000 cases yet to be clarified.

Another objective of the Kameradschaftsbund was to achieve the release of funds collected during the winter of 1944/45. At the time, soldiers of the Fallschirmpanzerkorps HG in East Prussia had collected 1.2 million Reichs Marks to support comrades and their families who were in dire circumstances due to the bombing and the escape to the west. The Kameradschaftsbund now wished, despite the devaluation, to use the funds deposited at the Berliner Sparkasse before the end of the war, for the intended purpose since misery and destitution still prevailed. Despite the fact that the fund had a definite purpose, the Senat of Berlin had confiscated it as 'military-nazi' property and planned to use it for its own ends. It required a long legal battle through a number of courts, unselfishly led by our comrade, Oberleutnant (ret.) Dr. Reimar Spitzbarth on behalf of the Kameradschaftbund, until the Kammergericht (superior court) in Berlin found that the Kameradsschaftbund was to be given the fund and the right to determine its use. Generalleutnant (ret.) Schmalz established the foundation "Kameradschaftshilfe" (help to comrades) which was able to support the poorest of the poor, mostly severely disabled by the war, for about 14 years.

The Kameradschftbund has furnished one of the lecture halls at the Kampftruppenschule of the Bundeswehr in Munster as a 'room of tradition' and was able to donate to the 'collection of artifacts of armored combat troops' at the same school, in addition to two exhibit cases, further memorabilia of the Fallschirmpanzerkorps HG, so that the memory of the Fallschirmpanzerkorps HG is being kept alive there as well.

At the museum of war history in Rastatt, a historical deed is documented. Conceived on the initiative of Oberfeldwebel (ret.) Walter Brusmeier and donated by the comrades, a display case recalls, in picture and words, the saving of the art treasures of the Benedictine monastery Monte Cassino by Oberstleutnant Schlegel of the Fallschirm-Panzer-Division HG.

Chronology of Development

February 23, 1933	Polizeiabteilung Wecke z.b.V. established in Berlin-Kreuzberg
June 1933	Enlarged to Polizeigruppe Wecke z.b.V.
July 17, 1933	With the creation of the Landespolizei, renamed Landes-polizeigruppe Wecke z.b.V.
January 12, 1934	Renamed Landespolizeigroup GG
September 23, 1935	Incorporated into the Luftwaffe as RGG
March 1, 1942	RGG enlarged to Regiment (mot.) HG
July 21, 1942	Enlarged to Brigade HG in France
October 17, 1943	Division HG created in southern France
May 21, 1942	After the destruction of the Division in north Africa; Re-establishment as Division (mot. trop.) HG, simultaneous creation of Brigade z.b.V. HG in southern France and Italy (Naples area)
July 15, 1943	Re-organization tp Panzer-Division HG in Sicily
January 6, 1944	Re-named Fallschirm-Panzer-Division HG
September 24, 1944	Establishment of Fallschirmpanzerkorps HG near Modlin: —General command and division strength Korps troops —Fallschirm-Panzer-Division 1 HG —Fallschirm-Panzergrenadier-Division 2 HG —Fallschirm-Panzer-Ersatz and Ausbildungs-Brigade HG (in Rippin/West Prussia)
March 14, 1945	Establishment of Fallschirm-Panzer-Ersatz and Ausbil-dungs-Brigade 2 HG in Velten and Joachimstal/Mark Brandenburg as replacement for the Fallschirm-Panzer-Ersatz and Ausbildungsbrigade HG which was lost at the fortress Graudenz.
May 9, 1945	Capitulation of the Wehrmacht and dissolution of all remaining units in Saxony, Berlin, at the Oder front, in Bavaria, Holland and the Rhineland. The majority of survivors became prisoners of war, mostly to the Soviets.

The Commanders

Unit	Office Holder		
	From	Rank	Name
Polizeiabteilung Wecke z.b.V.	23. Febr. 1933	Major der Schutzpolizei	Wecke
Polizeigruppe Wecke z.b.V.	Juni 1933	Oberstleutnant der Schutzpolizei	Wecke
Landespolizeigruppe Wecke	17. Juli 1933	Oberst der Landespolizei	Wecke
Landespolizeigruppe GG	12. Jan. 1934	Oberst der Landespolizei	Wecke
Landespolizeigruppe GG	6. Juni 1934	Oberstleutnant der Landespolizei	Jakoby
RGG	23. Sept. 1935	Oberstleutnant	Jakoby
RGG	22. Aug. 1936	Major i.G./Oberstleutnant/ Oberst	von Axthelm
RGG	1. Juni 1940	Oberst	Conrath
Verstärktes Regiment (mot.) HG	1. März 1942	Oberst	Conrath
Brigade HG	21. Juli 1942	Oberst	Conrath
Division HG	17. Okt. 1942	Oberst/Generalmajor	Conrath
Divisionsteile in Afrika	Jan. 1943	Oberst/Generalmajor	Schmid
Division (mot.trop.) HG	21. Mai 1943	Generalmajor	Conrath
Brigade z.b.V. HG	21. Mai 1943	Oberst	Schmalz

Panzer-Division HG	15. Juli 1943	Generalmajor/Generalleutnant Conrath
Fallschirm-Panzer-Division HG	6. Jan. 1944	Generalleutnant Conrath
Fallschirm-Panzer-Division HG	16. April 1944	Generalmajor Schmalz
Fallschirmpanzerkorps HG Kommandierender General	24. Sept. 1944	Generalmajor/Generalleutnant Schmalz
Fallschirm-Panzer-Division 1 HG	24. Sept. 1944	Oberst/Generalmajor von Necker
Fallschirm-Panzer-Division 1 HG	9. Febr. 1945	Oberst/Generalmajor Lemke
Fallschirm-Panzergrenadier- Division 2 HG	24. Sept. 1944	Oberst/Generalmajor Walther
Fallschirm-Panzer-Ersatz- und Ausbildungs-Brigade HG	24. Sept. 1944	Oberst Meyer
Fallschirm-Panzer-Ersatz- und Ausbildungs-Brigade 2 HG	14. März 1945	Oberst Breuer

The Holder's of the Knight's Cross

The following soldiers were awarded the Knight's Cross during their service with an HG unit. Rank at the time of award.

1. Febr. 1945	Oberst	Walther, Erich	als 131. Soldat

Holder of the Knight's Cross of the Iron cross with Oak Leaves with Swords

21. Aug. 1943	Gen.Maj.	Conrath, Paul	als 276. Soldat
23. Dez. 1943	Oberst	Schmalz, Wilhelm	als 358. Soldat
24. Juni 1944	Major	Fitz, Josef	als 511. Soldat
1. Febr. 1945	Major	Roßmann, Karl	als 725. Soldat
28. Febr. 1945	Obstlt. i.G.	von Baer, Bern	als 761. Soldat
15. April 1945	Major	Ostermeier, Hans	als 834. Soldat

Holders of the Knight's Cross of the Iron Cross with Oak Leaves

4. Sept. 1941	Oberst	Conrath, Paul
6. Okt. 1941	Oblt.	Graf, Rudolf
12. Nov. 1941	Oblt.	Roßmann, Karl
23. Nov. 1941	Lt.	Itzen, Dirk
7. Mai 1943	Obfw.	Schäfer, Heinrich
8. Mai 1943	Hptm.	Kiefer, Eduard
21. Mai 1943	Gen.Major	Schmid, Josef
21. Juni 1943	Hptm.	Schreiber, Kurt
21. Juni 1943	Obfw.	Scheid, Johannes

2. Aug. 1943	Major	Kluge, Waldemar
2. Aug. 1943	Hptm.	Rebholz, Robert
5. April 1944	Hptm.	Quednow, Fritz
5. April 1944	Lt.	Knaf, Walter
18. Mai 1944	Obgefr.	Witte, Heinrich
9. Juni 1944	Major	Hahm, Konstantin
24. Juni 1944	Oberst	von Necker, Hanns-Horst
25. Juni 1944	Oberst	von Heydebreck, Georg-Henning
5. Sept. 1944	Feldw.	Kulp, Karl
30. Sept. 1944	Hptm.	Bellinger, Hans-Joachim
30. Sept. 1944	Hptm.	Thor, Hans
6. Okt. 1944	Hptm.	Schmidt, Fritz-Wilhelm
10. Okt. 1944	Oblt.	Lehmann, Hans-Georg
18. Okt. 1944	Major	Sandrock, Hans
18. Okt. 1944	Hptm.	Stronk, Wolfram
19. Okt. 1944	Obfhr.	Birnbaum, Fritz
20. Okt. 1944	Hptm.	François, Edmund
29. Okt. 1944	Uffz.	Kalow, Siegfried
31. Okt. 1944	Oblt.	Schuster, Franz
20. Nov. 1944	Hptm.	Stuchlik, Werner
30. Nov. 1944	Oblt.	Kuhlwilm, Wilhelm
30. Nov. 1944	Oblt.	Kraus, Rupert
30. Nov. 1944	Lt.	Wallhäußer, Karl-Heinz
30. Nov. 1944	Uffz.	Grunhold, Werner
30. Nov. 1944	Gefr.	Plapper, Albert
30. Nov. 1944	Gefr.	Steets, Konrad
6. Dez. 1944	Hptm.	Renz, Joachim
6. Dez. 1944	Oblt.	Tschierschwitz, Gerhard
14. Jan. 1945	Major	Briegel, Hans
15. Jan. 1945	Lt.	von Majer, Hans
15. Jan. 1945	Feldw.	Köppen, Eckhard
28. Jan. 1945	Hptm.	Wimmer, Johann
7. Febr. 1945	Hptm.	Kampmann
8. Febr. 1945	Lt.	Koenig, Heinz
11. Febr. 1945	Hptm.	Hansen, Hans-Christian
18. Febr. 1945	Maj. i.G.	Schweim, Heinz-Herbert
19. Febr. 1945	Gefr.	Schirner, Lothar
23. Febr. 1945	Obfhr.	Hartelt, Wolfgang
.Febr. 1945	Obfw.	Herbst, Erhard
23. Febr. 1945	Gefr.	Rademann, Emil
28. Febr. 1945	Obgefr.	Krappmann, Heinrich
12. März 1945	Oblt.	Klein, Armin
26. März 1945	Lt.	Lippe, Hans
26. März 1945	Uffz.	Probst, Heinz
März 1945*	Oberst	Bertram, Eric
5. April 1945	Hptm.	Schulze-Ostwald
17. April 1945	Lt.	Leitenberger, Helmut
28. April 1945	Feldw.	Zander, Wolfgang
April 1945	Hptm.	Rommeis, Gerhard
9. Mai 1945	Oberst	Meyer, Friedrich
9. Mai 1945	Lt.	Behre, Friedrich

*probably on March 28, 1945

This picture of the simple grave of an unknown German soldier, is representative of all graves where one of our comrades found his last place to rest, unknown to us or no longer reachable. It is also to remind us of the unknown fate of one missing in action, whose family waited in vain for the return of their son, husband, father or brother. The grave site should also remind us of the many dead comrades whom we recall in honor.

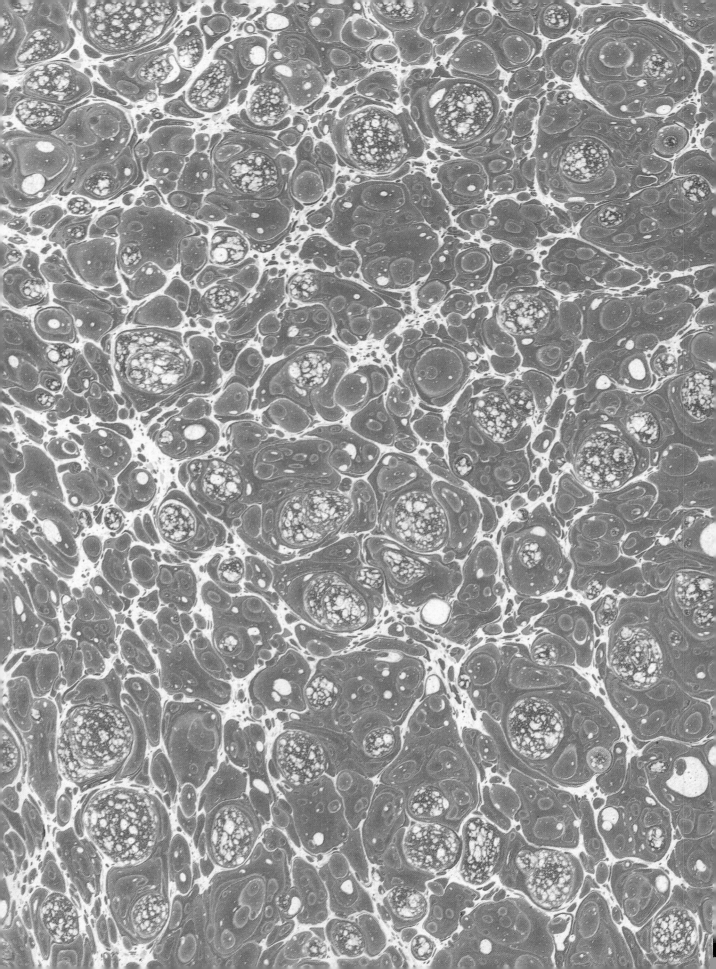